"Life always begins again. . ."

# THE
# FIRST
# ESSENE

*Six Days of Lectures by*
**EDMOND BORDEAUX SZEKELY**

*at the July 1979 International Seminar
on The Essene Way and Biogenic Living
held at the International Correspondence Center
of the International Biogenic Society, Orosi, Costa Rica*

**MCMLXXXI
INTERNATIONAL BIOGENIC SOCIETY**

## SOME BOOKS BY EDMOND BORDEAUX SZEKELY

THE ESSENE WAY—BIOGENIC LIVING
THE ESSENE GOSPEL OF PEACE, BOOK ONE
BOOK TWO, THE UNKNOWN BOOKS OF THE ESSENES
BOOK THREE, LOST SCROLLS OF THE ESSENE BROTHERHOOD
THE DISCOVERY OF THE ESSENE GOSPEL: The Essenes & the Vatican
SEARCH FOR THE AGELESS, in Three Volumes
THE ESSENE BOOK OF CREATION
THE ESSENE JESUS
THE ESSENE BOOK OF ASHA
THE ZEND AVESTA OF ZARATHUSTRA
ARCHEOSOPHY, A NEW SCIENCE
THE ESSENE ORIGINS OF CHRISTIANITY
THE ESSENE TEACHINGS FROM ENOCH TO THE DEAD SEA SCROLLS
THE ESSENES, BY JOSEPHUS
THE ESSENE TEACHINGS OF ZARATHUSTRA
THE ESSENE SCIENCE OF LIFE
THE ESSENE CODE OF LIFE
THE ESSENE SCIENCE OF FASTING
ESSENE COMMUNIONS WITH THE INFINITE
THE FIRST ESSENE
COSMOTHERAPY OF THE ESSENES
THE LIVING BUDDHA
TOWARD THE CONQUEST OF THE INNER COSMOS
JOURNEY THROUGH A THOUSAND MEDITATIONS
FATHER, GIVE US ANOTHER CHANCE
THE ECOLOGICAL HEALTH GARDEN, THE BOOK OF SURVIVAL
THE TENDER TOUCH
MAN IN THE COSMIC OCEAN
THE DIALECTICAL METHOD OF THINKING
THE EVOLUTION OF HUMAN THOUGHT
THE GREATNESS IN THE SMALLNESS
THE SOUL OF ANCIENT MEXICO
THE NEW FIRE
DEATH OF THE NEW WORLD
ANCIENT AMERICA—PARADISE LOST
PILGRIM OF THE HIMALAYAS
MESSENGERS FROM ANCIENT CIVILIZATIONS
SEXUAL HARMONY
LUDWIG VAN BEETHOVEN, PROMETHEUS OF THE MODERN WORLD
BOOKS, OUR ETERNAL COMPANIONS
THE FIERY CHARIOTS
CREATIVE WORK: KARMA YOGA
THE ART OF STUDY: THE SORBONNE METHOD
COSMOS, MAN AND SOCIETY
I CAME BACK TOMORROW
THE BOOK OF LIVING FOODS
CREATIVE EXERCISES FOR HEALTH AND BEAUTY
SCIENTIFIC VEGETARIANISM
THE CONQUEST OF DEATH
HEALING WATERS
THE BOOK OF HERBS, VITAMINS, MINERALS

3/96

*Book Design & Typesetting by Golondrina Graphics*

# PREFACE

I arrived to Costa Rica for the first time on July 24, 1979, though I had been planning my trip for many months before. It was in November of 1978 that I had written to Dr. Szekely, asking if I could visit him at the new international correspondence center of the International Biogenic Society and, as a Teacher of Biogenic Living, perhaps help in some way, knowing that the projected International Seminars would need planning and preparation. I was delighted to receive his answer—"Wonderful idea!"—and made my own preparations to leave my home in the Pacific Northwest for the sunny climate of Costa Rica.

Two days after I arrived, the first *International Seminar on the Essene Way and Biogenic Living* to be held in Costa Rica, began. Previous Seminars had taken place, for many years, mostly in California. The group was small, and this proved to be a blessing, for although Professor (as Dr. Szekely was most often called) always enjoyed a Socratic dialogue regardless of the size of his audience, he was undoubtedly more inspired than usual by the enthusiastic personal interaction and participation made possible by a small group.

Watching Professor personally demonstrate the ancient Sacred Ball Game of precolumbian America, or dramatically enact the role of Ahura Mazda during the 8000-year-old Sumerian "Creation of the Universe," none of us imagined that just thirteen days after the close of the last session he would die. Even now, it is hard to think of him gone, and I can still hear his deep laughter and see his eyes, so full of humor and sparkle. We are very grateful that tapes were made of this, his last Seminar, so that we who were fortunate enough to attend might be able to share the experience with thousands of readers all over the world. We are glad, too, that in the transcription of the tapes, Professor's unique way of expressing himself was left completely untouched—reading the words somehow makes it live for us again.

The title of this book may seem strange to those who do not know of Edmond Bordeaux Szekely's intense involvement with the Essenes and their writings for the last fifty years. In view that he translated the *Essene Gospel of Peace* from ancient Aramaic, wrote about the Dead Sea Scrolls before they were discovered, and brought the attention of the world to the teachings of the mysterious Brotherhood of the Essenes, the only "practical mystics in history," it seemed that no one in this century deserves more to be called "The First Essene." Knowing Professor, he would probably modestly disagree, preferring to refer to himself, in the spirit of his "old friend Socrates," as simply a midwife helping the truth to be born; what gave him the most pleasure was knowing he had stimulated an "intellectual fermentation" in the minds of his readers and listeners. But all of us who are part of the International Biogenic Society will gladly accept the responsibility of conferring on him this richly-deserved title—knowing that all those whose lives have been indelibly touched and changed forever by his books, will allow us to do so.

*Orosi, Costa Rica, the 9th of July, 1980.*

DANIEL J. SHEEHAN

*Come with us to lovely Orosi, in the summer of 1979. Feel the warm sunshine on your face, as you sit in the semi-outdoor lecture hall, waiting for Professor to step out the door, take his seat and say, in his "unforgettable baritone," "Well, well, well! What questions have you for me today?" Listen to the singing of the birds, little messengers of St. Francis from the nearby 300-year-old mission church. See the soft mist gather over the mountains to the southeast, heralds of the rainclouds to come in the afternoon. Let the colors and fragrance of the tropical flowers fill your heart with joy, and quicken your response to these eternal teachings you will hear—from one who not only spoke about them, and wrote about them, but who lived them—every day of his long life.*

## CONTENTS

*Note to the Reader: Questions and comments by Norma are so identified. All other questions asked of Dr. Szekely by the Seminar participants are in italics and scattered throughout the text.*

*Professor and Norma*

## The First Day: July 26, 1979

Very little, really, is known about Precolumbian Wisdom—for evident reasons, because the discovery of the American continent came late in history, and also because the discovery was made mainly by Spaniards who came in search of gold and riches, and not for the spiritual content of the existing cultures in America. Another reason is that descriptive archeology is not adequate to deal with the content of Precolumbian Wisdom.

I want you to realize that there is an independent parallel between the evolution of philosophical ideas in the old world, and on the American continent. For instance, the Mayan culture had great similarities to the culture of ancient Egypt. Both had similar pyramids, similar dress, similar dedication to astronomy and the calculation of time. Although there was no real connection between ancient Egypt and the Mayas, it is amazing how many similarities they have. There is also a great similarity between the Toltec culture and the classic Greek culture. We may say that all of western culture is nothing else but the variation, permutation and combination of the basic bricks laid down by the classic Greek genius—in philosophy, literature, the arts, and practically everything else. Everything came from them, first through the Romans and then through the Renaissance, where the different elements of classic Greek culture were renewed. In the same way, it was the Toltecs who created the great arts, philosophy, and literature on the American continent. And just as the Romans borrowed everything—literature, art, philosophy, mythology—from the Greeks, in the same way the Aztecs borrowed everything from the Toltecs. Also, the Romans and the Aztecs were both conquerors. In their time, the Romans conquered practically the whole of the known world, and in the same way, the Aztecs conquered all those other nations in the area of what is now Mexico and the northern part of Central America. The Aztecs had very few original ideas—everything they had, even their mythology, came from the Toltecs. So there is an interesting parallel in historical evolution, between the Egyptians, the Greeks, and the Roman culture in the old world, and the line of evolution between the Mayas, the Toltecs, and the Aztecs, in the new world.

Pyramids played an important role in both worlds, but their functions were very different. The pyramids of ancient Egypt, for

9

instance, were dedicated to the Cult of Death. They were mainly burial places for the Pharaohs. In contrast, the Toltec and Mayan pyramids were dedicated to the Cult of Life. There they created spiritual and teaching centers, where the youth were trained to become priests. They were also scientific centers in the study of astronomy and calculation of time. In fact, the Mayas had more accurate calendars and more punctual calculation of time, than existed in Europe at the time of the Conquest. They also served as centers of agriculture, advising people when to sow and when to harvest, and many other functions related to daily life. There are more pyramids in the area of Mexico than there are in Egypt, and they are also incomparably more interesting in their aspect as an integral function of society.

They had a dualistic concept of the universe, conceiving it as the Battle of Two Opposing Forces: the god Quetzalcoatl, the source of all good things for man, and the opposite god, Tezcatlipoca, the god of all things which are hindering man on this planet. The symbol of Quetzalcoatl was the feathered snake, and the symbol of Tezcatlipoca, the jaguar.* Because they believed that everything in existence was the result of the battle of opposing forces, their cosmic concept was dynamic, as well as dualistic. And the way they tried to express this concept resulted in the construction of one of the greatest mysteries in the realm of archeology: the Sacred Cosmic Ball Game of Precolumbian America.

The reason why archeologists have always been confused in trying to decipher the ruins of these ball game places, which they finally just concluded was some kind of ancient sport, is that they forgot to study the ancient codices, which hold the key to understanding, not only the meaning of the precolumbian Ball Game, but every other aspect of life in ancient America, as well.

The ancient codices are composed of native drawings—not texts, but pictographs. Pictographs consist mainly of physiograms and ideograms. A physiogram is a picture taken from nature. For instance, when we see a picture of a few drops falling down, it is evident it is rain. We think immediately of the sun when we see a circle with rays going to every direction, and so on. They expressed concrete ideas from nature with physiograms, and abstract ideas in the form of symbols, and ideograms.

*For a complete description of the archeology and philosophy of ancient America, please read *The Soul of Ancient Mexico, Ancient America: Paradise Lost, Death of the New World, The New Fire,* and *The Greatness in the Smallness,* all by Edmond Bordeaux Szekely, published by the International Biogenic Society.

QUETZALCOATL AND TEZCATLIPOCA

(from *El Juego de los Dioses*, by Edmond Bordeaux Szekely)

There are in existence about three dozen of these codices, in various museums and libraries in different parts of the world. The most important one is the Florentine Codex, which I used to illustrate my books *Ancient America: Paradise Lost,* and *Death of the New World.* This codex is really an encyclopedia of life in precolumbian America—there you will find all the plants of ancient America, all the animals, all the arts, all the customs, all the little industries, even a history of mythology and many other things. We owe the existence of this codex and many others to a very interesting Franciscan monk who came to the American continent with Cortez. Fran Bernardino de Sahagun (who also plays an important role in my book *The New Fire*) was as intelligent as he was spiritually dedicated. A gifted linguist, he learned over twenty native languages, and with the instinct of an archeologist he collected tremendous numbers of native artifacts, sculpture, paintings, drawings—which represented every facet of native life in ancient America. He virtually dedicated his life to this work, and eventually compiled his famous *Historia General de las Cosas de Nueva España,* a twelve-volume anthropological, mythological, social, and natural history of ancient Mexico. His enthusiasm for his task led him into trouble, however. In spite of the fact that he was a Franciscan monk, he was accused of fomenting idolatry, condemned by fanatical orthodox elements and ordered to destroy his whole collection, his entire life's work. Well, he didn't like very much this idea, and after meditating and praying for guidance a brilliant plan came to his mind. He mobilized all his native friends, made about fifty large boxes, packed them with all his treasures of art and literature, and addressed them to the King of Spain as a gift to the throne. Of course, nobody had the courage to destroy property of the King, so in this way he saved his collection, although they did have to be shipped to Spain, as that was where the King was. When the boxes arrived to the Spanish court, the King and his friends had an excellent time with these exotic things for several months—of course, not understanding one word of what they really were—but finally they got bored with them, and at the advice of one whom we would call today a ʼ ʼ ːr of Culture, sent them as a present to the faculty of phil and archeology at the University of Florence, in Italy, at th one of the greatest centers of culture in Europe. There, ulty of philosophy and archeology made pictures of ever and

12

created a collection which is called today the Florentine Codex. History, art, literature, social life, philosophy, crafts, botany, zoology, all is contained in the pages of the Florentine Codex, and in fact, Sahagun describes in his many volumes of "The General History of the Things of New Spain" only what is in the Florentine Codex, which were all his collections, sent to the King of Spain in order to save them.

In view of this, we have a complete bankruptcy in descriptive archeology regarding precolumbian culture, because the soul, the spirit, the quintessence of it is in the codices, with their vivid pictographs. All this immense treasure-house of information was neglected by descriptive archeologists because they base everything only on the ruins they find and the artifacts they excavate, and ignore this vital, basic source of information, the ancient codices.

This is why the ruins of the ball game places have been generally ignored and classified simply as some strange precolumbian sport. If they were to study the codices, they would realize that these ball game places were really the reproduction of the universe in miniature. Each one was a microcosm where were collected all the forces of the universe, all the elements of nature, and the structure and function of the universe, all in a small area. I was so impressed by the philosophical and spiritual implications of this "game" that I reconstructed several, the most recent one here, and also one in the state of Jalisco, Mexico, for the University of Guadalajara, at the request of the founder of the University, and a very good friend of mine, Lic. José Guadalupe Zuno (also the father-in-law of the former President of Mexico, Luis Echeverría). He was really dedicated to my discoveries in precolumbian Mexico and to what he called the Cosmic Ball Game. I also reconstructed at my retreat called Mille Meditations, eighty miles east of San Diego, a ball game which was played at the time of King Netzahualcoyotl (which means "hungry coyote" in Nahuatl). He was called the "King Solomon" of ancient Mexico and richly deserved the title, being a great poet, philosopher, astronomer, and sage. He also built a magnificent ball game place which I reconstructed at Mille Meditations in order to teach precolumbian wisdom.*

*A complete description of King Netzahualcoyotl's version of the Sacred Cosmic Ball Game, as well as an analysis of the philosophical meaning of the various types of the Ball Game, as played all over what is now Mexico and Central America, may be found in *Journey Through a Thousand Meditations,* by Edmond Bordeaux Szekely, available from the International Biogenic Society.

13

The Cosmic Ball Game is not really a game, but a microcosmos—a complete microcosmos in miniature of everything which is going on in the universe. And for the purpose of explanation, this version which I reconstructed here at the I.B.S. International Correspondence Center, played by the Huetar Indians, will be much simpler and less involved than the Toltec Ball Game, and therefore easier to teach, so you will understand the meaning.

This simple Central American version needs six persons, so let us see who wants to play it. . . first, who is willing to become the god Quetzalcoatl? Fine, Dan. Now, who wants to be Tezcatlipoca, the jaguar god, the opposite? We must have one person who is willing to be the lord of darkness! Fine, David, very good. Now I need someone to be the Priest of the Drums. Good, Norman. And I need someone else to be Priest of the Numbers. Fine, Henry. Now, who wants to be King of the Moon—no, I am afraid there cannot be a Queen of the Moon, if we are faithful to history—I am innocent! Fine, we have a King of the Moon, and now we need a King of the Stars. Good, now I have my six persons. Now we will go down and I will explain to you the whole microcosmos, the quintessence of precolumbian wisdom, and we will actually play the game, because when we want to bring an ancient teaching to life, there is nothing better than putting it into practice. You remember the Chinese proverb, that one illustration is better than a thousand words. *(The entire class adjourns to the area of the precolumbian ball game.)*

The basis of this microcosmos is this symbol, shaped like an X: these four straight lines going to the center—this is the symbol of Ollin, a sacred symbol shared by all precolumbian cultures. The Mayas, Toltecs, Aztecs, Miztecs, Chichimecs, all had this universal symbol. What does it represent? First, the four cardinal points: south, north, east, west. It represents also the four seasons of the year: spring, summer, fall, and winter. It represents the four elements: air, water, earth, fire. And in the center of this symbol is the Sun. The stone ring on top of the column is also called the sunstone, because the sun is the cause of the four cardinal points, and the sun is the cause of the four seasons of the year. So you have the sun in the center, and you have the sacred symbol of Ollin, representing the four cardinal points, the four seasons of the year, and the four elements.

Now into this microcosmos come the players: the one represen-

*Quetzalcoatl scores a goal and ushers in a million years for the Kingdom of Light!*

ting Quetzalcoatl and the kingdom of light, and the one who is Tezcatlipoca, and the kingdom of darkness. Each of them must stay at this point *(a square stone some distance from the sunstone)* and cannot move during the game. And especially they must not step on their symbol directly in front of them—for example, the player representing Quetzalcoatl must not step on the sacred symbol of the feathered snake, otherwise he is disqualified and his opposite wins the game by default. So behave.

Now here is the ball, which Quetzalcoatl and Tezcatlipoca will try to throw through the ring of the sun. The ball represents the planet earth, and that is very interesting, because the ancient Toltecs, Aztecs, etc., had no idea that the earth was round, that it was a globe. They chose a round ball to represent earth purely by instinct. The Toltecs used one filled with hard rubber—*ule*—from a tree in Mexico with the same name. Solid rubber is very heavy, and it must have been very difficult even to lift such a ball, let alone throw it, but they were much stronger than we are. Probably they didn't eat white sugar, white flour, and white rice! The home of Quetzalcoatl is the Evening Star—here is the star. And the home of Tezcatlipoca, the jaguar, is the moon. The jaguar goes to hunt at night, and the moon represents darkness.

So here we have the four cardinal points, the four seasons, the four elements, there we have the sun, here we have the earth, there we have the stars, there we have the moon—this is the precolumbian microcosmos. Everything they had to know about the universe is here.

On this long wall we have the precolumbian numbers, starting from each end and going to the center. One point represents one pitaya, a small round desert cactus fruit, two pitayas are two, three pitayas are three, four pitayas are four. Five is represented by a piece of sugar cane. So one sugar cane and one pitaya is six, one sugar cane and two pitayas are seven, and so on. Two sugar canes are ten, and the number nineteen would be three sugar canes and four pitayas. And the last is the sacred number of precolumbian wisdom: twenty. The symbol of twenty is the open sea shell. We have two cubes, one with the picture of Quetzalcoatl, and one with a picture of Tezcatlipoca, and when the Cosmic Ball Game starts, the High Priest of the numbers puts one cube at one end of the wall, and the other cube at the other end of the wall, each on number one. Every time Quetzalcoatl makes a goal, the Priest of

Professor guides the players through the mysteries of the Precolumbian Ball Game. The ancient precolumbian numbers are painted on the low wall.

the Numbers moves the cube to the next number, and each time Tezcatlipoca succeeds with a goal, he moves the opposite cube. And the Priest of the Drum rings the bell (because we could not find a drum) every time the ball goes through the stone of the sun. Whoever reaches the center first, won the game. Now a very important thing: let us say that Quetzalcoatl is throwing the ball, and it falls down. Then the King of the Stars grabs it quickly and throws it back to Quetzalcoatl. Or if the ball falls down in the triangular area of the moon, then the King of the Moon quickly grabs it and throws it to Tezcatlipoca. Because the ball must always be in continuous motion, just as the earth is always in continuous motion around the sun.

*(The game is played, with a resounding victory for Quetzalcoatl and the Kingdom of Light!)*

Well, well! Now that Quetzalcoatl triumphed over Tezcatlipoca, everything will be all right with the universe!

Now I want to explain to you another part of the cosmic ball game. This small, one-room construction just the other side of the wall of the numbers, is called a *Temazcalli*. In every village in precolumbian America there were always a few Temazcalli, which required no plumbing to provide a perfect way of taking a bath. Outside is a little fireplace where they put the wood and made a fire, and inside is a small area where water was boiling, and that water created vapor, and there the player sat down to perspire and recuperate from the effort required to make twenty goals. All the impurities were cleansed from his system, and this is one of the reasons the natives had such perfect skin, which was praised by the Spaniards and written about in their reports.

So the Temazcalli was a very healthy idea, but it also had a deep symbolic and philosophical meaning. The walls were made of adobe, earth, that was one element; the fire burning was another element; the vapors going up, that was the air; and the water boiling was another element. So in the Temazcalli were all the four elements, represented by the four gods: the god of fire, Tonatiuh, who was also the god of the Sun; the god of air, who was also the god of the wind, Quetzalcoatl; the god of earth, Chimalma, and the god of rain, or water, Tlaloc. It was a kind of communion, relaxation, meditation, surrounded by the four gods.

And just as the Temazcalli was simply, physically a vapor bath, but spiritually and philosophically it was a communion with the

THERE WAS A PLAYER WHO REPRESENTED QUETZALCOATL . . .

(from *El Juego de los Dioses*, by Edmond Bordeaux Szekely)

four gods, the four elements, so the Cosmic Ball Game was in one sense an athletic game, but at the same time it was an all-sided microcosmos, representing the whole universe. Everything had a material function parallel with a spiritual function.

*(The class returns to the outdoor lecture hall.)*

Now I want to call your attention to another aspect of this Cosmic Ball Game—not the philosophical or symbolical aspect, but what it does to the human body. I was always interested to know why historians such as Antonio de Solis, Bernal Díaz del Castillo, Fray Bernardino de Sahagun, etc., all described the tremendous physical power of the Aztecs they first encountered. For instance, when later ships arrived, often a very heavy box would have to be carried by four Spanish soldiers with great effort—and suddenly came one Aztec, put the box calmly on his shoulder and beautifully walked away with it! Well, one of the reasons for this physical strength was the playing of the Cosmic Ball Game.

Just imagine yourself throwing the ball: first the foot muscles work, then the lower leg muscles, then the upper leg muscles, the diagonal abdominal muscles, the thoracic muscles, and also the biceps and the muscles of the fingers. It is a good example of the precolumbian genius to create a game which is not only a unique spiritual and philosophical synthesis, but also an athletic exercise which sets into movement all the muscles of the human body in right proportion.

On this subject, there is an interesting intermezzo in the history of the Aztecs which I think you will enjoy. Montezuma, the emperor, liked very much fresh fish from the ocean for his dinner. But the Aztec capital, Tenochtitlan, now Mexico City, is a good distance from the ocean, as you well know. So he set up a system distributing several runners between the ocean and the capital, so one would catch the fish early in the morning, run to the second one, the second took over when the first probably collapsed, then to the third, to the fourth, and so on, so the fish could arrive for his dinner. It was not a very democratic country; nevertheless, the emperor was the emperor, and a very obstinate person. It was a beautiful idea, but it didn't work, because the distances were just too great, and the runners always collapsed before the fish was able to reach its destination. So he started to meditate again, and got the idea to use his best warriors. He put them in equal distances between the ocean and the capital and waited for his fish. It was a disaster. The best warriors were efficient when it came to killing the maximum amount of people in a battle, but nevertheless they collapsed too, and the emperor still didn't have his fish. So finally a priest advised him to use the Ball Game Players. The emperor was willing to try anything in order to have his fresh fish—he was addicted to fresh fish—so they made a relay team of the Ball Game Players, and they did the job beautifully. Minutes before the sun was setting, they were able to place before the emperor the fresh fish which that morning was still in the ocean.

This story tells something about the physical fitness of the Ball Game Players. They turned out to have more endurance than the runners, more endurance than the soldiers, more endurance than anybody else the emperor tried. You see, it is a mistake *not* to live in the 20th century, but it is an even greater mistake to live *only* in the 20th century, because we can learn from ancient cultures and ancient civilizations many things which can enrich our lives in the present, and be very useful for our health and happiness. Now a little Ball Game Place like this one is only 16 yards by 8 yards—or meters, because in archeology we work with meters. You can build one on any small lot, and even the construction is very good exercise. And the playing of the game is much better for the body than jogging, or whatever exercise fad is currently in style. It not only develops the muscles in the right proportion, but tremendously improves the circulation. When you improve your circulation you improve the distribution of the vitamins—because they are distributed by the blood—you improve the distribution of minerals—because the circulation of the blood carries minerals to the cells—you improve the distribution of enzymes, for the same reasons, and you also improve the distribution of the hormones. And when the distribution of vitamins, minerals, enzymes, and hormones are improved, your general health is also greatly improved. You will perspire, you will get rid of catabolic waste products, your circulation will be improved, and you will have relaxation at the same time. You don't have to believe in the mythology of the four gods of the precolumbian Indians—Quetzalcoatl, Tonatiuh, Tlaloc, and Chimalma—but it will do something wonderful for your health. And I also want to remind you that it is not a sauna, but a vapor bath, which is hot, but not too hot. The dry heat of the sauna is too intense for many people, and those with high blood pressure who take a sauna may end in 48 hours in the cemetery. Also, too much heat is not good for someone with nervous tension. But the vapor bath, the Temazcalli, is a gentle procedure, which everyone, even with high blood pressure or nervous tension, can benefit from, because the temperature never goes that high. And the first thing you will notice is the effect on the skin. All those authors I mentioned wrote about the Indians, describing the perfect color and texture of their skin, comparing it to the pale, weak skin of the Spaniards.

You ask about the sunstone—it is 2 meters high. That is the only thing which is not so easy to build, but it can be done. When I constructed a Toltec Ball Game (which is much more complicated) for the University of Guadalajara in Mexico, I found a fine stone carver, gave him my books on ancient Mexico, and he copied the symbols and made a beautiful sunstone, one which weighed about a ton when he was finished. I told my masons to build a very strong stone column to support the weight, and they did. And then arrived the day when the sunstone had to be lifted onto the column. My masons were all young Indians, very intelligent, working with great efficiency. I asked their foreman, a talented young man called Juanito, if I should arrange with a company in Guadalajara to bring out some heavy machinery to lift the sunstone—the reconstruction was at my home on Lake Chapala, about fifty miles from Guadalajara—he gave me an offended look, saying, you don't trust us? I told him of course I trusted him, but the stone weighed a ton and I didn't see how he could lift it to the top. He laughed and told me to sit down and watch, adding, "you forget that our ancestors built the pyramids!" I asked him if he would bet me he would not be able to lift it up, and he replied that he never bet on anything he knew he would win. So with that, I just settled down comfortably with a good book and waited to see what would happen. Well, I saw them picking up all kinds of funny little sticks, pieces of wood, rope, bits of metal, and they started to move things, levers with ropes, etc., and to my great surprise, slowly, slowly it started to go higher, ever so gradually. It took a good number of hours, but at the end they put the stone on top of the column! It was unbelievable, but I saw it! They had not only patience, but ingenuity, too, and the phylogenetic heredity. Their ancestors really did build the pyramids, and there must have been some similar technique.

Also, I want you to realize that in the world today there are only about 12% of the population who have normal plumbing. Most of mankind lives in remote areas everywhere in poverty, without adequate plumbing or hygiene. In Asia, in Africa, in parts of Latin America, people who live in remote areas, people who have absolutely no way to install plumbing or bathrooms, could benefit greatly from this vapor bath of the Aztecs. This is something practical which could be done for hundreds of millions of people in Asia, in Africa, in Latin America, and not only improve their health and hygiene, but also cost practically nothing. And it is good for the ecology. The only fuel necessary is a handful of wood, and one quart of water, which anyone can afford. It doesn't use energy, it doesn't interfere with the ecology—it is an excellent example of my philosophy that we can learn a lot from ancient civilizations, and renew these wonderful things to improve our health and the quality of life, without destroying the environment or going to a lot of expense which most of mankind cannot afford.

Regarding precolumbian America, generally speaking, there are two schools in archeology: the Americanist and the Hispanist. The Hispanists maintain that everything good which was brought to ancient America was brought by the Spaniards—a common language, a superior religion, and many other things. That is one concept. The other concept is put forward by the Americanist school, which maintains that the original cultures of America were superior to the culture of the conquerors, and all the Spaniards did was to plunder and destroy all the great works of art and literature of ancient America in their quest for gold, and by forcing a new religion on the people they did much more harm than good. I belong to the Americanist school, maintaining that we' had a superior civilization in ancient America, and invariably at archeological conventions I would be fighting with my colleagues about this subject. So finally I decided to settle this problem once and forever, by writing a book—*Death of the New World*. I selected from all the dozens of codices, from the National Museums, from private collections of Diego Rivera and other important ones, vivid drawings of natives who were present at the arrival of the Spaniards, who actually saw everything that happened, and who sent to the Aztec emperor messages as the conquerors came closer and closer to the capital, warning him about these strange creatures who had all kinds of instruments which vomited fire, and who

Professor explains the philosophy and use of the *temazcalli*—precolumbian vapor bath. *(below)* The Cosmic Battle of Light and Darkness in progress.

rode monsters with four legs (because horses were unknown in that part of the world). These drawings, so graphically describing the conquest, virtually day by day, were facts, not archeological speculation. They were not made by historians two or three hundred years later, but by people on the scene, eye-witnesses to one of the major events of our planet's history. So in this book, as well as another one I wrote on the same subject, *Ancient America: Paradise Lost,* I have hundreds of native drawings which show, not only all the details of daily life in ancient America, but also exactly how the Spaniards massacred the Indians and brought death to the new world.

There are also original Nahuatl texts—small literary masterpieces which describe poignantly all that happened to them during the conquest. Here is one which is particularly graphic:

> "Now we cease to be a people;
> Shall no longer be a nation.
> All of us are disappearing
> And our knowledge is departing;
> All our arts and artifices;
> All the wisdom of our people;
> All the glory of our nation.
> We have brought destruction on us.
> No salvation is there for us.
> Woe to us, O race fill-fortuned!
> May our hearts be filled with courage
> To confront the coming evil!

In order to understand what it must have been like for them, we must try to imagine our feelings if some space ship were to arrive, disembarking inhuman monsters with unknown weapons who would suddenly attack us and destroy everything we had known and believed in, including ourselves. It must have been a similar inner and outer experience for the natives of ancient America. Compared to the natives, the Spaniards had far fewer soldiers, and the natives were in much better physical condition. Nevertheless, they could not fight the bullets and cannons, and they were terrified of the horses, because it looked to them as if the horse and man were one, creating a new kind of monster which demoralized them completely.

However, although the conquest of the capital, Tenochtitlan, took only a few years, there were a few setbacks for the Spaniards, and the study of one of them led to an interesting adventure for me. There was a struggle in a place called Aguascalientes, and Cortez sent his main army to occupy the area. They set up camp at night, as usual, in order to make a final attack the next morning. But during the night the whole Spanish army was annihilated! Only three or four were able to escape and give a report about what happened. And their report only added confusion to what is still one of the mysteries of the history of the conquest: how was it possible that with watchmen everywhere, and unquestioned military and technical superiority, that they were surprised so completely, losing a battle in such a definitive way that it set back the conquest by twenty years? I was always puzzled by it—how it was possible for a primitive people to destroy a strong army with excellent soldiers with guns, cannons, all the weapons of war, who had watchmen everywhere to make sure they were never surprised? So I decided to go myself to Aguascalientes and look around, and I discovered some very interesting things.

It seems that the capital of the state of Aguascalientes (which is also called Aguascalientes) is built over a system of underground caves. There is a veritable cobweb of ancient tunnels threading their way under every part of the city, and on the walls of the tunnels there are all kinds of ancient pictographs—practically all with the same motif—a huge bird diving downwards. Even today there remain legends of gigantic birds swooping down from the sky and attacking animals and people. And not only legends, but giant eagles still exist in that area, probably descendants of those early, very aggressive ones, which do dive on their prey of small animals and carry them off.

Making a closer study, I discovered that the native population in that long-ago time, in their fear of the attacks of the giant birds, made these underground tunnels and caves as a place of refuge and escape. They felt a kind of security, knowing they had somewhere to go if the birds would suddenly swoop down on them. When those early tribes died out, the tunnels remained, and they were already very old at the time of the conquest.

And then it struck me what must have happened that night when the Spaniards suffered their defeat. They set up their camp, secure in the safety of their perimeters, and the Indians simply went into the tunnels and came up during the night in the middle of the Spanish camp. The one place they were not watching, the very center of their encampment, was the site of the massacre. It took me two years to figure out what happened, and it set back the conquest by twenty years. And even that twenty years refers only to the center of the empire, the area around the capital. There were many, many outposts still so far away that the natives still did not know about the conquest. It was this fact, together with the archeological reconstruction of one of ancient America's most beautiful and meaningful religious rites, that led to the writing of my book, *The New Fire.*

In the precolumbian cultures, one generation was calculated as fifty-two years. As each generation drew to an end, all old things had to be destroyed: old clothes, old utensils in the kitchen, old tools, and most important of all, old fires. In anticipation of a new cycle, it was important that all existing fires of the empire be extinguished, and no one was allowed to keep a fire. It was a very interesting concept of the cycles of the universe. Then four high priests were chosen, representing the four gods of the four elements—Quetzalcoatl, Tlaloc, Chimalma, and Tonatiuh— and these four priests of air, water, earth, and fire went to the top of the highest mountain and started a New Fire. The actual moment when the sticks ignited was holy, for it was thought that Ometecuhtli, the God of gods, had forgiven man his sins and would not destroy the world but allow a new cycle of life to begin.

The best runners of the empire were standing by, and as soon as the New Fire was ignited, they immediately lit their torches from it and ran to the four cardinal points according to the symbol of Ollin—they could not deviate from that—and carried the fire to every corner of the empire. This was the great ritual of the Tonalamatl, and the celebration of it just ten years after the conquest of Tonichtitlan caused a series of events which inspired my book, *The New Fire.*

It seems that one of Cortez' captains decided to make himself King of Guatemala, and Cortez had to send his army to put down the rebellion. The capital remained without an army and without protection, and it was just at this time that a generation, the fifty-second year, came to a close. And all the tribes who lived far away from the capital who had not yet heard about the conquest, and were therefore not yet conquered, began to converge on the capital to celebrate the Tonalamatl, the creation of the New Fire. So they came, and the Spaniards saw to their great consternation thousands of Indians pouring in to the capital through the night, from the valleys, mountains, and orchards. They had no idea they were coming for a religious ceremony—that they did not even know of the presence of the Spaniards—and of course they were sure it was the beginning of an uprising. So there was a panic, because the army had been sent away to Guatemala to subdue that rebel captain. That is the background of my book, *The New Fire,* which describes these happenings in the form of a drama, and gives a good account of the confrontation of the two cultures. Something few people are aware of is that precolumbian America was not really conquered by the Spanish army, but by the poor Franciscan monks who came with them—barefoot, often in rags, they helped the Indians, cured their diseases, and taught them about agriculture and many useful things to help them live better. They helped them as fellow human beings in need, not as slaves and subjects. Theirs was the true spiritual conquest, and without it the material one could never have taken place. If you read *The New Fire,* you will be able to understand much better precolumbian wisdom, and how the followers of King Netzahualcoyotl and the followers of St. Francis avoided a holocaust through a harmonious synthesis of the spiritual beauty of their two worlds.

You ask about Costa Rica, and why it was never conquered. Well, let us look at Mexico first. The Aztecs were a centralized kingdom—the Spaniards came, captured the emperor, captured the capital, and everything collapsed. The Incas were a centralized kingdom—Pizarro came with his armies, captured the emperor, captured the capital, everything collapsed. The weakness of central power, of centralization, is one of the most interesting lessons of history. We shall think about this and learn for the twentieth century. (This is why I am a decentralist!) Now in Costa Rica, there was no central power, only about a dozen Indian tribes scattered over the country. So when the Spaniards arrived, they did not find an emperor, nor a central capital, nor any kind of organized central power. And this is why the conquest of the Aztecs and the Incas took only about five years, and why the conquest of Costa Rica took about fifty years. Although the Indians were not able to defeat the Spaniards militarily, they moved farther and farther back into the mountains, destroying their crops, their corn, their *pejivalle* trees, etc., as they retreated. And the Spaniards were unable to defeat them. It was a similar situation as with Napoleon in 1812 when he attacked Russia—the scorched earth policy—destroy everything and retire to the mountains. It is a perfect example of the great strength of a decentralized society over a centralized one. It reminds me of what the Roman emperor, Caligula, said once: I wish all my enemies could have one head so I could chop it off.

This defeat of a stronger enemy through decentralization happened only here, in Costa Rica. And the consequence was, of course, that the Spaniards couldn't make slaves out of the Indians to build them large haciendas, as they did in Mexico, or put thousands of Indians to work in the mines. The mighty conquerors had to get down from their horses and cultivate the land themselves in order to survive. There is a beautiful phrase in Isaiah: And they shall beat their swords into plowshares, and their spears into pruning-hooks. . . well, this is exactly what the Spaniards had to do. It was hard work, but they had no slaves to do it for them—they had to do it themselves if they wanted to survive.

But here is an interesting thing: no one can cultivate a large area by himself. Using the primitive methods of that time, no one person could alone cultivate more than an acre of land. Therefore, a society of smallholders developed, and in this kind of society the system of government that works the best is democracy. And this was the origin of Costa Rican democracy, and why it is so deeply imbedded in the hearts and minds of the Costa Ricans. Those original smallholders, with their independent minds and genius for self-sufficiency, are the roots of the democratic society that now flourishes in Costa Rica. That tree bears abundant fruit today—free elections, free press, civil rights, a very liberal constitution which is honored and followed—all this in a tiny country which is an oasis of freedom surrounded by dictatorships.

By the way, this is why there are not too many of you at this Seminar, because so many participants were afraid of civil war in Costa Rica, and switched their reservations to December. Of course, they forgot that the war was in Nicaragua, not in Costa Rica! Costa Rica has not one soldier, nor munitions factories, and the country is just as peaceful now as it was when the trouble began in Nicaragua. But it is interesting to see how many people know so little about geography. We have a proverb in France which says that a Frenchman is a little moustache who doesn't know the geography. Well, I don't have a moustache, and though

I am French, I think I know a little more about geography than most people who are always getting the countries in Central America confused, when they are in reality so different, one from the other.

So this is why we received letters and calls from people, mainly in the United States, asking if we were still alive, and if the bombs were falling, and hoping the massacres would not cause us trouble. So they all switched to the December Seminar. But I am delighted with a small group, as it is always easier to answer questions and progress with the material. And the material I will discuss with you tomorrow is the most important of the Seminar, and something difficult to discuss and demonstrate with one or two hundred people. So please have a good biogenic feast, and I will see you all tomorrow!

## The Second Day: July 27, 1979

Two thousand years ago, the Roman governor Pontius Pilate asked, what is truth? Well, we have several thousand religions at the present in the world, and most of them maintain that they are in possession of the only and exclusive truth, and all the others are wrong. And I, as a philosopher, suspect that they are correct, because probably they all are wrong—for the simple reason that whenever someone believes he is in possession of the only exclusive truth and everything else is wrong, he has committed the greatest mistake in life: one-sidedness. We may say that person or organization has committed a kind of intellectual suicide—they have built a stone wall between themselves and the rest of the Cosmic Ocean and will not search anymore.

We have a general confusion in the world at present, because we live in an age of disorientation, no question about it. We never had such a mess on our planet as we have today, from every viewpoint. In the field of philosophy, we are not much better. It was about fifty years ago when I was in England to give my annual summer seminar that my British publisher ushered me into his orchard, locked the gate in a most friendly way, and told me that unless I would write a certain little book I had promised him a long time before, he would not let me out of the orchard. Well, I enjoyed the orchard but I still had important business in London, so I had to sit down and write it for him. This book is now called *The Evolution of Human Thought,* in which I wrote about 100 philosophers and 80 philosophical schools, giving about ten lines to each philosopher and school. Today, after fifty years, we are much worse. We have not 100 philosophers, but probably 500 philosophers, and the question is, are any of them able to answer that question of Pontius Pilate?

What is truth. . . Fortunately, we have an excellent yardstick, coming to us from the most ancient culture in history, Sumeria, by the most ancient philosopher we know, Zoroaster the first, or Zarathustra. His statement is about eight thousand years old, and he says this: If we want to know whether or not something is true, we must check it with three measurements: (1) Is it in harmony with the *laws of nature,* or not? (2) Is it in harmony with your *inner intuition*—does it have for you the power of evidence? (3) Is it in harmony with all the great teachings of the past, with *ageless wisdom?*

31

If something is in harmony with only one of the three, it is not enough, or with only two of the three, then you had better continue to search, until you find something which is in harmony with all the three: with the laws of nature, with your inner intuition, and with ageless wisdom. I have yet to find a better guide in life than these three criteria. You cannot go wrong if what you adopt as your philosophy is in harmony with these three yardsticks. So let us analyze each of the three.

Regarding the laws of nature, a lot of people are accustomed to the phrase in a superficial way, using expressions and notions for which we have words, but not really knowing what is behind the words. The laws of nature were never defined so thoroughly, so precisely, nor with greater clarity and simplicity, than in a conversation held about two thousand years ago between Josephus Flavius, the Roman historian, and Banus, an Essene teacher. Josephus Flavius spent several years in the Essene Brotherhood at the Dead Sea, where he was assigned a master called Banus. In the form of questions and answers they beautifully defined what are the laws of nature. Incidentally, the original manuscript of their conversations was discovered unexpectedly by a French philosopher, Count Volney, a follower of Voltaire, in an attic in Alexandria, apparently owing its survival to the Copts who carried it there. I translated and published it under the title, *The Essene Code of Life.* * Before we go further, I want you to listen to this beautiful conversation which defines with the utmost clarity the Laws of Nature.

THE TEN ESSENTIALS OF THE NATURAL AND COSMIC LAWS

*Josephus:* What are the essentials of the natural and cosmic laws?
*Banus:* There can be assigned ten principal ones.
*Josephus:* Which is the first?
*Banus:* To be inherent to the existence of things, and consequently, primitive and anterior to every other law: so that all those which man has received, are only imitations of it, and their perfection is ascertained by the resemblance they bear to this primordial model.
*Josephus:* Which is the second?
*Banus:* To be derived immediately from the Creator, and presented by him to each man, whereas all other laws are presented to us by men, who may be either deceived or deceivers.

---

*The Essene Code of Life,* as well as all the other books by Edmond Bordeaux Szekely mentioned in this book, are available from the International Biogenic Society, mailing address: I.B.S. Internacional, Apartado 372, Cartago, Costa Rica, Central America.

*Josephus:* Which is the third?

*Banus:* To be common to all times, and to all countries, that is to say, one and universal.

*Josephus:* Is no other law universal?

*Banus:* No: for no other is agreeable or applicable to all the people of the earth; they are all local and accidental, originating from circumstances of places and of persons; so that if such a man had not existed, or such an event happened, such a law would never have been enacted.

*Josephus:* Which is the fourth essential?

*Banus:* To be uniform and invariable.

*Josephus:* Is no other law uniform and invariable?

*Banus:* No: for what is good and virtue according to one, is evil and vice according to another; and what one and the same law approves of at one time, it often condemns at another.

*Josephus:* Which is the fifth essential?

*Banus:* To be evident and palpable, because it consists entirely of facts incessantly present to the senses, and to demonstration.

*Josephus:* Are not other laws evident?

*Banus:* No: for they are founded on past and doubtful facts, on equivocal and suspicious testimonies, and on proofs inaccessible to the senses.

*Josephus:* Which is the sixth essential?

*Banus:* To be reasonable, because its precepts and entire doctrine are conformable to reason, and to the human understanding.

*Josephus:* Is no other law reasonable?

*Banus:* No: for all are in contradiction to reason and the understanding of men, and tyrannically impose on him a blind and impracticable belief.

*Josephus:* Which is the seventh essential?

*Banus:* To be just, because in that law, the penalties are proportionate to the infractions.

*Josephus:* Are not other laws just?

*Banus:* No: for they often exceed bounds, either in rewarding deserts, or in punishing delinquencies, and consider as meritorious or criminal, null or indifferent actions.

*Josephus:* Which is the eighth essential?

*Banus:* To be pacific and tolerant, because in the law of nature, all men being brothers and equal in rights, it recommends to them only peace and tolerance, even for errors.

*Josephus:* Are not other laws pacific?

*Banus:* No: for all preach dissension, discord and war, and divide mankind by exclusive pretensions of truth and domination.

*Josephus:* Which is the ninth essential?

*Banus:* To be equally beneficent to all men, in teaching them the true means of becoming better and happier.

*Josephus:* Are not other laws beneficent likewise?

*Banus:* No: for none of them teach the real means of attaining happiness: all are confined to pernicious or futile practices; and this is evident from facts, since after so many laws, so many religions, so many legislators and prophets, men are still as unhappy and ignorant as they were thousands of years ago.

*Josephus:* Which is the tenth essential of the natural and cosmic laws?

*Banus:* That it is alone sufficient to render men happier and better, because it comprises all that is good and useful in other laws, either civil or religious, that is to say, it constitutes essentially the moral part of them; so that if other laws were divested of it, they would be reduced to chimerical and imaginary opinions devoid of any practical utility.

*Josephus:* Recapitulate all ten essentials.

*Banus:* We have said that the law of nature is:

1. Primordial
2. Immediate
3. Universal
4. Invariable
5. Evident
6. Reasonable
7. Just
8. Pacific
9. Beneficent
10. Alone Sufficient

And such is the power of all these attributes of perfection and truth, that when in their disputes the theologians can agree upon no article of belief, they recur to the natural and cosmic laws, the neglect of which, they say, forced God to send from time to time prophets to proclaim new laws; as if God enacted laws for particular circumstances, as men do; especially when the first subsists in such force, that we may assert it to have been at all times and in all countries the rule of conscience for every man of sense or understanding.

These ten definitions of the natural laws, the laws of nature, I couldn't find anywhere more completely described than these few words. So always please try to remember these ten criteria: Primordial, Immediate, Universal, Invariable, Evident, Reasonable, Just, Pacific, Beneficent, and Alone Sufficient. Whatever satisfies these ten requirements is a law of nature, a natural law. Well, there is a simple example from history: when an apple fell on Newton's nose in the orchard, he discovered the law of gravity. And there are many other things—if you plant a seed, and there is humidity, sunshine, the preconditions of life, a plant will grow. We are surrounded by the manifestations of the laws of nature, and this is a very important criterium. This is the most immediate thing for us—we don't have to look into extremely abstract complex and complicated concepts, because Zarathustra also said that if we learn to read from the Book of Nature, we can learn everything. He was the same philosopher who said that the noblest of all professions is to be the gardener of the earth. In fact, to teach the laws of nature, he used his orchard. That was his school. There he was every day with his disciples, and he illustrated everything he taught with an example from the garden and orchard. Nature is the primordial source of knowledge, and one of the three great verifications of truth.

Now let us analyze the second yardstick: intuition. All the great thinkers of different ages came to their ideas through a kind of intuition which religious people like to call revelation, and which in Oriental philosophy is called illumination. The hallmark of intuition is that it affects us usually with the power of evidence. Examples are the best way to understand something, so I will pick up the example of Buddha, the greatest philosopher of ancient India. As a young man, Prince Siddhartha, he was surrounded by all the riches of the royal court. But he felt an emptiness, a kind of intuition that this was not real life, this was not the truth, that there was something more. He felt a deep intuition to search, and so he started on his quest for knowledge. He tried many things, visited a great number of teachers and philosophers, but he could not find what he was looking for. And then one day, thinking under a mango tree, there suddenly came to him a great impact of an intuition. He suddenly realized that we are living in and are surrounded by an Ocean of Suffering. He realized that the world is full of violence, persecution, intolerance, ignorance, greed. . . he looked around and saw a lot of sick people, a lot of old, decrepit people; he saw people fighting with each other, hurting each other, deceiving each other. And he wrote down these immortal words: Life is suffering. To be born is suffering. To become ill is suffering. To become old is suffering. To be away from those we love is suffering. To be together with people we don't love is suffering. Not to achieve what we want to achieve is suffering. To achieve it but to lose it later is also suffering. Therefore, life is suffering. This deep intuition hit him with elementary force, and did lead him to the Four Noble Truths based on this single intuition. They called it in the language of the time, Samadhi—Illumination. He said the most important thing in life we must learn is to understand suffering, the Truth of Suffering. Then we must understand the Cause of Suffering. Then we must understand the Path Leading Toward the Cessation of Suffering. And, last, we must understand the Cessation of Suffering. These are the Four Noble Truths of Buddha. All are derived from this basic intuition of suffering.

Then he asked, what is the reason of suffering? According to Buddha, there are two reasons for suffering: one he called *Avidja,* ignorance; the other he called *Tanha,* thirst—thirst for all harmful pleasures, thirst for excessive material possessions, or greed, thirst for all those things which we don't need. Avidja and Tanha, ignorance and thirst, are pulling us through a labyrinth, an endless ocean of suffering. And if we want to get out from this ocean of suffering, then we have to eliminate ignorance and thirst. The philosophy of Buddha is extremely clear and simple. So clear and

THE WHEEL OF LIFE OF
# BUDDHA

(from *The Living Buddha*, by Edmond Bordeaux Szekely)

simple that Buddha had more followers than any religion on our planet. And the whole thing started with that basic intuition of suffering, and changed the lives of hundreds of millions of people. Intuition acts with the power of evidence, and there is no greater example of this than Buddha.

Now we have a third measurement to check if something is true or not. This is ageless wisdom. From time to time in history a great genius appears and establishes a sound philosophy of living which affects the lives of great numbers of people. Then time goes on, there come the followers, the master passes away, and the followers add all kinds of things—they dress this beautiful, simple, original teaching with dogmas and rituals, until with time it becomes more and more static, and finally loses its primeval power. Then after centuries and centuries another great genius appears with another great teaching, and the same thing happens again. It seems that the history of human thought and beliefs is a kind of pilgrimage from altar to altar through thousands of years. Now if, at present, I could call all these great founders of religions to a round table discussion, I guarantee you that they would agree on every essential point. But if I would call to a round table discussion all their disciples and followers, I guarantee you they would disagree on every point. Therefore, the best thing is to recognize the ageless wisdom of the greatest teachers and the greatest sacred books of mankind. The original ones—like the *Zend Avesta* of Zarathustra, the *Vedas* of ancient India, the *Tripitaka* of Buddha, the *Tao Te Ching* of Lao Tzu, the *Essene Gospel of Peace,* these are great ageless values which teach the same things, with the same profundity, the same evidence, the same clarity. There-fore, we must avoid one thing: to read and study the commentaries of the commentaries—yes, that's right, because this is exactly what happens: we have these beautiful, ageless, sacred writings. Then centuries and millennia pass, the commentaries accumulate, and then appear the commentaries of the commentaries, then the commentaries of the commentaries of the commentaries—and finally we have a sea of obscurity and confusion. The only way is to go back to the original purity and simplicity of these great teachings. The original purity and simplicity always stand the test of time. And they are there, through millennia, through thousands and thousands of years, we still are reading these great teachings in their original form.

If you look around today, you will see that every day thousands of books appear. Thousands of mediocre books which are forgotten in ten years and practically all of them forgotten in a hundred years—no one will remember for long the best-sellers of today. But the great pillars of knowledge: the *Zend Avesta,* the *Vedas,* the *Upanishads,* the *Teachings of Buddha,* the *Tao Te Ching* of Lao Tzu, the *Essene Gospel of Peace*—these are eternal values, which will still be eternal values a hundred thousand years from today. Therefore, if you want to know if something is true or not, you shall verify it, not only with the laws of nature, not only with your own intuition, but also with eight thousand years of wisdom. If all the three: the laws of nature, your inner intuition, and eight thousand years of wisdom, point to the same thing, then you cannot go wrong. You will know you are on the right path. And this path was right eight thousand years ago, and will be right a hundred thousand years from today. It is extremely important to realize this.

Another example of ageless wisdom, one closer to twentieth century man and easier for you to understand, is the philosophy of Epicurus. You probably have read his name here and there, and the commentaries of the commentaries told you he was a kind of hedonistic Greek philosopher who advocated the pleasures of life. Well, nothing is farther from the truth. He did advocate pleasures, yes. But he divided pleasures into two categories: the noble pleasures, and the harmful pleasures. What are these harmful pleasures, which he rejected outright? They are those for which we have to pay a very high price: the sacrifice of our physical health and our peace of mind. And there is no pleasure in the world which is worth our physical health and our peace of mind. All those pleasures which everyone is pursuing in this age, all fall into this category.

Then he described a second category of pleasures, what he called the noble pleasures—like the enjoyment of all beautiful things in nature: a beautiful sunrise, sunset, the ocean, the mountains, the sky full of beautiful clouds, travel, love, good books, good music— these are the noble pleasures of Epicurus. He called them our Eternal Companions. According to Epicurus, his wisdom and philosophy consist of a practical program of living, gradually replacing all our wrong pleasures with the right ones. He is also an excellent example for ageless wisdom.

The reason I brought up Epicurus as an example from ancient Greek philosophy, is that it was the Greek genius which laid down the foundation for our whole western culture. Everything we are doing is nothing else but the variation, combination, and permutation of the basic building bricks laid down by the classic Greek genius. For instance, Greek philosophy is the foundation of our entire history of philosophy. Greek sculpture is the foundation of a beautiful art, renewed by the Renaissance and carried on to the present age. Painting, also, is a beautiful art, renewed by the Renaissance. Their system of law was adopted by the Romans, and survives in the form of many laws today. Their theatre—Aeschylus, Sophocles, Euripides—was the foundation of all theatre and drama for thousands of years. Also for thousands of years the foundation of our arithmetic and geometry was provided for by geniuses such as Pythagoras and Archimedes. We can always go back to the ancient Greeks and find ageless wisdom.

What is the difference between wisdom and knowledge? Knowledge can be very impressive, it is true. But you may be a walking British encyclopedia and know everything, and still you can make a mess of your life, even with all your knowledge. Although knowledge is very useful, and in fact I am a great admirer of knowledge in every field, still the fact is that knowledge is not wisdom. I admit it may be very interesting to make a detailed study of the functions of the neural system of insects, but it is not necessary for our happiness. According to the great philosophers, wisdom is the sum total of knowledge we need for our happiness.

Now you have three excellent yardsticks in your hands to verify if something is true or not. Pontius Pilate did not have the means to answer his own question—what is truth—but you can answer it if you use these three yardsticks: the laws of nature, which are primeval, eternal and universal; intuition, which acts with the power of evidence; and eight thousand years of wisdom—the best of the values created by the human mind and the greatest geniuses of history. Therefore, when we will start to discuss the different teachings which are the subject of this seminar, please try yourself, actively, always to compare them to see if they are in harmony with these three yardsticks.

As I told you, we live in an age of disorientation and confusion. I have had an opportunity to see what is going on in the twentieth century until now, and comparing it with other centuries, I never found so much confusion than in this century. And *historia est vitae magister:* history is the teacher of life. We can learn from the errors of previous cultures. For example, if we understand the errors of the Roman Empire, we may avoid those mistakes they committed and so avoid their fate.

We are living now in an exaggerated technological civilization. We created such complexities of life that man's nervous system is gradually disintegrating under these complexities of life. If you look around, you will see nothing but the violations of these three yardsticks to truth, everywhere. You see chaos in every field. We reached the age when an unbelievable number of thermonuclear weapons are stored in many hands. All that is needed is that some-one handling these weapons would suddenly have high blood pressure, or a strong disharmonious emotion, and the button may be pushed and there we all go up in smoke, together with the masterpieces of Michelangelo in the Sistine Chapel, all the symphonies of Beethoven, the wonderful works of Bach and Mozart, the plays of Shakespeare, the writings of Goethe, everything which is great, all lost forever. And we arrived to this point by neglecting these three yardsticks.

We have now entered into an era of self-exploitation. Yes, you may be surprised by that expression: self-exploitation. You learned a lot about exploitation of man by man, but there is one form of exploitation which is the worst kind: self-exploitation. What is it? When we sacrifice the real values of life—good health, free time to enjoy the great masterpieces of literature, music and art, and everything which is beautiful in nature—for things we don't need, which are superfluous, or even harmful. We sacrifice real values to make a lot of money, and then we buy a lot of things which destroy our health and peace of mind. Also, never before in history has there existed in such excess the production of unnecessary and harmful things, as in the present day. Go anywhere, to a supermarket or department store, you will see what I mean.

It reminds me of one of my favorite old friends, a philosopher from ancient Greece called Diogenes. Diogenes was a strange person. He wanted to impress his contemporaries with the need to stop self-exploitation and a complicated life, so he was living in a barrel! Yes, he was taking a daily sunbath beside his barrel; when it was raining, he went under the barrel, turning it upside down; then when he wanted to walk farther, he rolled his barrel toward another orchard, picked up some figs and oranges to eat, went a little farther, found a goat, milked the goat and had some milk, and so on. But all these eccentricities served a purpose: to impress his fellow man about the great truth in a simple life. He went to an extreme in order to teach them how to live simply.

One day, his friends persuaded him to go with them to the famous market-place of Athens, saying that he didn't know what he was missing, describing the fantastic array of merchandise there from fifty countries, etc., etc. So, finally, in order to get rid of them he agreed to accompany them to the market-place. And they spent the whole day there and saw everything. At the end of the day they asked him, well, now you have to recognize that all these things are truly marvelous, and you must admit that you are depriving yourself by not enjoying some of these wonderful things. He had a very simple answer for them: "I never realized there were so many things in the world which I don't need!" And it is much more true today. Wherever you go, in whatever city, you will find 99% of the things which you don't need.

This can be applied also in the field of intellectual life and philosophy, and in the printing of books. Never in my life in any previous age did I see so many mediocre books and literature—not only mediocre, but definitely harmful. It is incredible that we are cutting down our trees and destroying our ecology in order to print thousands and thousands of books, totally without value.

I will always remember the words of a wonderful professor I had when I was studying at the University of Paris at the Sorbonne. Once we were walking by the side of the Seine, past all the book stalls selling good books, bad books, mediocre books, masterpieces, everything, an ocean of books. And he said to me, "My son, remember that each time you read a mediocre book, you sacrifice a great masterpiece, because time is not unlimited, and you could read a great masterpiece instead of reading that book—but you may never have the time to read that great masterpiece when you wasted it reading a mediocre book."

I suggest you read one of my books, with the title, *Books, Our Eternal Companions.* There I write about my greatest friends, books, in all their aspects—books and religion, books and philosophy, books and freedom, books and democracy, and at the end, I give a list of a hundred books which I consider are the greatest masterpieces of universal literature. It is a very fine list to use in order to set up priorities to follow when you study.

I never saw so many people reading so many mediocre books than at this period of the twentieth century. In previous ages, we had fewer books, but they were great books. We shall try to extricate ourselves from an ocean of mediocre literature, from the commentaries of the commentaries of the commentaries, and go back to great books.

Lucretius said, *cognoscere est cognoscere causas.* To know is to know the causes, the beginning, the reasons for things. It will be impossible for us to understand something if we don't know its origin. Therefore, when we want to start studying the great teachings, we have to study their origin first.

Lucretius said that if we want to understand the world and the universe, we must understand its beginning. This is why all the sacred books of mankind—the *Zend Avesta* of Zarathustra, the *Rig Veda* of ancient India, the *Tao Te Ching* of Lao Tzu—all try to understand the Creation, the Beginning of the Beginnings, the Origin of the Universe.

Therefore, we will explore a little bit these ancient writings, going chronologically inverse, back to the origin. I want you to know what the *Tao Te Ching* says about the Creation, what the *Rig Veda* says, what the *Zend Avesta* of Zarathustra says. And we will start with the *Tao Te Ching,* a short quotation dealing with the beginning of the beginnings, by Lao Tzu, the author of the *Tao Te Ching,* a little book which has never been surpassed in condensing so much truth in such a small place. It is truly an example of the Greatness in the Smallness! Here is what he said about the beginning of the beginnings:

In the beginning of heaven and earth
There were no words;
Words came out of the womb of matter.
And whether a man dispassionately sees
To the core of life,
Or passionately sees the surface,
The core and the surface are essentially the same,
Words making them seem different,
Only to express appearance.
If name be needed, wonder names them both:
From wonder into wonder, existence opens.

Something mysteriously formed,
Born before heaven and earth.
In the silence and the void,
Standing alone and unchanging,
Ever present and in motion.
I do not know its name.
Call it Tao.

Man follows the earth.
Earth follows heaven.
Heaven follows the Tao.
Tao follows what is natural.

This is from the *Tao Te Ching,* the very deep concept of ancient Chinese philosophy. Of course, Lao Tzu doesn't tell you concretely the secret of the origin of the origins because a definition is by nature limited, and reality can never be compressed into a little package of definitions, but it will appeal to your intuition. Also, it is interesting that he mentioned what is natural: "Tao follows what is natural." Tao was his concept of the Cosmic Order, a tradition which arrived to China from ancient Sumeria.

Now let us see another great ancient sacred book, the *Vedas,* the foundation of the philosophy and wisdom of ancient India. This is the Hymn of the Creation from the most ancient of the *Vedas,* the *Rig Veda.*

Then was not non-existent nor existent: there was no realm of air, no sky beyond it.

What covered in, and where? And what gave shelter? Was water there, unfathomed depth of water?

Death was not then nor was there aught immortal: no sign was there, the days and nights divided.

That one thing, breathless, breathed by its own nature: apart from it was nothing whatsoever.

Darkness there was: at first concealed in darkness, this All was indiscriminated chaos.

All that existed then was void and formless: by the great power of warmth was born that unit.

Thereafter rose desire in the beginning, Desire, the primal seed and germ of spirit.

Sages who searched with their hearts' thought discovered the existent's kinship with the non-existent.

Transversely was their severing line extended: what
was above it then, and what below it?

There were begetters, there were mighty forces, free
action here and energy up yonder.

Who verily knows and who can here declare it, whence
it was born and whence comes this creation?

The gods are later than this world's production. Who
knows, then, whence it first came into being?

He, the first origin of this creation, whether he formed
it all or did not form it,

Whose eye controls this world in highest heaven, he
verily knows it—or perhaps he knows not.

This is the struggle of the Indian mind, the genius of ancient
India, with the least-definable thing, the Creation, the beginning
of the Cosmos. This is thousands of years older than the *Tao Te
Ching,* and now we will go even farther back in history, to the
*Zend Avesta* of Zarathustra, over eight thousand years old. It is
the most ancient sacred book of mankind—it goes farther back
than ancient Egypt, or China, or India. The *Zend Avesta* is the
first encyclopedia of mankind, the first sacred book which really
represents the limit in chronology we can go with the existence of
alphabets—it is a kind of sign on the highway as we go back in
time—behind that we will meet only pictographs.

In this part of the *Zend Avesta,* Zarathustra asks Ahura Mazda,
the Creator, about this same subject, the origin of the universe.
This will give you a good idea of the style of the *Zend Avesta,* as
the most ancient of law-givers, Zarathustra, tries to find out from
Ahura Mazda the mysteries of the Creation and the Beginning:

Who, by generation, was the first Father
Of the Cosmic Order within the world?
Who gave the recurring Sun and Stars
Their undeviating way?
Who established that whereby the Moon doth wax
And whereby she waneth?
Who from beneath supported the Earth
And, from above, hath upheld the Clouds,
That they do not fall?
Who hath made the Waters, and who maketh the Plants?
Who to the Wind hath yoked the Storm-Clouds,
The swift and even the fleetest?

Who, O Creat Creator! Is the Inspirer of
The Good Thoughts within our Souls?
Who, as a skillful Artisan,
Hath made the Light and the Darkness?
Who, as thus skillful, hath made Sleep
And the zest of the waking hours?
Who spread the Auroras?
The noontides and the midnight?
For whom hath Thou made the Mother-Earth,
The producer of joy?
This do I ask of Thee, O Ahura!
Pray tell me it rightly: That Holy Faith
Which is, of all things, the Best,
And which, going on hand in hand with Thy people,
Shall further my lands in ASHA, Thine Order.

Before we go further, I want to mention a few things about this first encyclopedia of mankind, the *Zend Avesta*. It has parts which deal with the movement of the celestial bodies, with the universe. It has chapters which deal with the elements on our planet—the sun, the air, the water, the earth. There are chapters dealing with trees, with vegetation. If you think organic gardening is something newly discovered, you will be greatly surprised when you read the *Zend Avesta*. It is not organic gardening which is new, but the other kind of gardening, which uses chemicals. For eight thousand years, man used the natural kind of gardening.

The *Zend Avesta* deals not only with every element, with all the plants, trees, foods, fruits, vegetables, but also with man—it deals with the influence on man by these different fruits and vegetables and the elements of our planet. Philosophy, poetry, art—all these too are in the *Zend Avesta*. And in spite of its scope, it is not like the British Encyclopedia, because that set of books contains a tremendous amount of knowledge, while the *Zend Avesta* contains a tremendous amount of wisdom, that wisdom which is essential for our happiness.

The interesting thing is that the *Zend Avesta* is itself a recapitulation of previous ancient traditions, lost in the mist of history. These early heliolithic teachings were written only in pictographs, in pictographic language. What are pictographs? You may think that pictographs are not as good as written language, based on alphabets. Well, not necessarily. If you analyze our alphabets and our written language, you will realize that it is all abstract. I don't think that letter A or letter B grow on trees. I never had the opportunity to shake hands with letter H or to meet letter O walking on the sidewalk. Our letters of the alphabet are nothing more than disjointed abstractions—they have no reality. They are rather substitutes for reality.

When pictographs deal with things of nature—for instance, if there is a drawing of a few drops of water falling down, it doesn't matter who will read it, whether British or German or French or Russian, they all will immediately understand that it stands for rain. A drawing of a circle with rays going out from it is instantly understood by everyone to mean the sun. These are physiograms, images taken from nature. Those which express an idea are ideograms. Yesterday, during our study of precolumbian wisdom, we discussed the symbol of Ollin, a symbol of movement, representing the four cardinal points, the four seasons, the four elements—this is an ideogram, representing an idea. But as we go farther and farther back we will meet less abstraction and more reality in the pictographs. Our alphabetical language, which we use to combine letters into words and sentences, is just a dry skeleton of reality, while pictographs represent the muscles, the bloodstream, the more intensive reality. This is why certain parts of the *Zend Avesta* are written in pictographs. In fact, in my English translation of the *Zend Avesta** I have in the appendix the original pictographs—physiograms and ideograms, and you can compare the English translation with the original pictographic version by reading them simultaneously. You will see that the Creation of the Universe, and especially the creation of the Earth, cannot be explained adequately only with our written language—we have to go back to pictographs. Zarathustra did it, and we have to follow him in this path. But never think for a minute that pictographs are inferior to our alphabetic language, because they are superior.

ASHA is the word Zarathustra used in the *Zend Avesta* for the Cosmic Order. This is the greatest aspect of the teaching of Zarathustra—the Cosmic Order—that we are not living in a capricious universe—that everything is in balance, in harmony, and it is the Cosmic Order which permits us to understand the universe

---

*The Zend Avesta of Zarathustra,* by Edmond Bordeaux Szekely, available from the International Biogenic Society.

and the cosmic laws, and to live according to them. This is the best of all things. In the ancient language of the *Zend Avesta,* ASHEM VOHU—Asha, the Cosmic Order, is the best of all things.

Now these are the questions asked by Lao Tzu, by the *Rig Veda,* by Zarathustra, the most ancient of our sages in history. This is the end of the path—we cannot go any farther back, as this is the most ancient of the teachings. So now we will start to explore the greatest mystery of the universe, the Creation.

First we will deal with the origin of the cosmos, of the universe. These are the symbols of the origin of the universe which the ancient Sumerians used, and which Zarathustra is giving us in the *Zend Avesta.* According to the Sumerian concept, everything started from a point. It may seem strange, and I don't want to appear blasphemous, but they considered that the Creator manifested in a point—a point has no width, no length, no thickness, a point is not in space, a point is not in time, but is omnipresent, everywhere. Everything is composed of points—we may call them atoms, or much smaller particles, but whatever it is called, it is the infinitesimal small point. According to their concept, at a time so remote it is impossible to conceive with the human mind, billion times billion times billion times of eons ago, there was a cosmic explosion of a point—and this explosion created Time, Force, Space, and Matter. Simultaneously there was a tremendous explosion which created the Cosmos.

Now we shall not be anthropocentric and think that Zarathustra speaks about our planet only, because our planet is an extremely small point in the solar system; the solar system is an extremely small point in our galactic system; and our galactic system is an extremely small point in the ultra-galactic system; and the ultra-galactic system is an extremely small point in the known universe, and the known universe is an extremely small point in the unknown universe. Therefore, we have to abandon megalomania, and not think our planet is in the center of the universe, because there are billions of planets even in our galaxy.

So, the creation of the universe is the tremendous cosmic explosion of a point, creating time, force, space, and matter. This is the ancient Sumerian concept of the beginning of the Creation, and it is very interesting that western science, physicist after physicist for a long time, was trying to give us a scientific explanation of the universe. There were all kinds of theories, including the Kant-Laplace theory, the theory of relativity, and many other theories, until finally came the most interesting of all: the Big Bang theory. After all those thousands of books and research and collection of data, they reached the conclusion that at a certain immense distance of billions of eons there was a big bang, and everything started to rush out in the universe. It is very interesting: there is western science, climbing a kind of Mt. Everest of theory after theory, going closer and closer to the top, through all these different theories, from Kant-Laplace to Einstein and the Big Bang, and finally our scientists arrive to the peak and there is good old Zarathustra sitting and smiling at them and saying, "Welcome!" Finally they reached the same concept as Zarathustra did eight thousand years ago in the *Zend Avesta*.

The first pictograph of mankind, then, was the point. It was the symbol of the beginning, out of which everything came. And this first, most ancient "Big Bang" was the basic, primeval Power, the creative principle, the point—suddenly moving and creating the Universe.

The first act of the Creator was to create Time, symbolized by a straight perpendicular line moving upward from the point. The next act of the Creator was the perpendicular line moving downward from the point, representing Force, or Energy. The two lines, one above the point and one below it, represent Time and Force together, the movement of energy in time, or the measurement of force or speed.

In the next act of creation, the point moves to the right in a horizontal line, and this signified Space. The Creator next moved in a straight line to the left from the point, symbolizing Matter. The two horizontal lines representing Space and Matter produced Volume, or Mass.

When the four lines above, below, to the left and to the right of the point were combined into a pictograph, it signified the creating of the universe; in other words, Time, Space, Force and Matter before the appearance of all their modalities. We may say it represents the cosmic nebulae, before the appearance of the solar systems and planets.

When the perpendicular line above the point, symbolizing Time, was multiplied by eight, four lines to the right of the perpendicular line, and four lines to the left of it, these lines represented Time divided into the eight seasons of the year, classified by Zarathustra

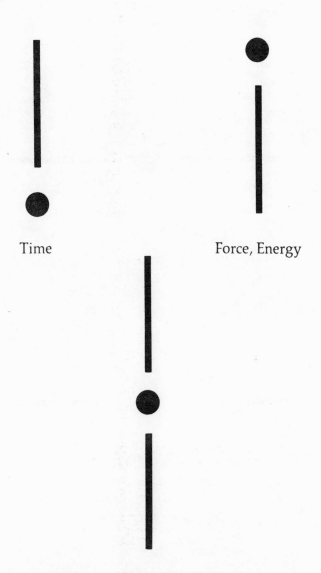

Creator, Creative Principle

Time

Force, Energy

Time and Force: Measurement of Force or Speed,
Movement of Energy in Time

Space

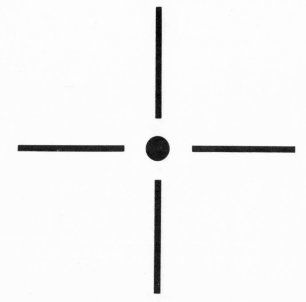

Matter

Space and Matter: Volume or Mass

Creating of the Universe: Time, Space, Force and
Matter before Creation of their Modalities

Eight Seasons: Spring, Spring-Summer, Summer,
Summer-Fall, Fall, Fall-Winter, Winter,
Winter-Spring

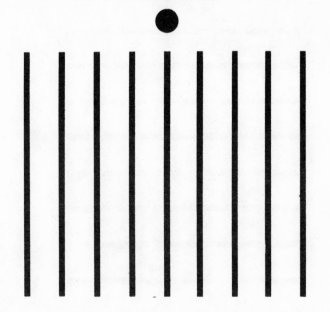

Good and Evil Energies from Stars, Sun and
Earth and those inherent in Man

Eight Cardinal Points: East, South-East, South, South-West, West, North-West, North, North-East

Good and Evil Forms of the Four Elements: Air, Water, Earth, Fire

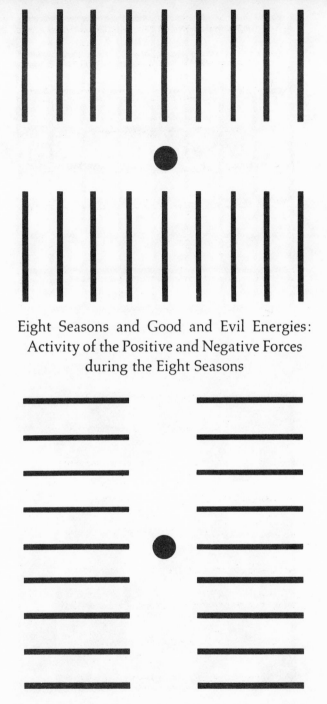

Eight Seasons and Good and Evil Energies:
Activity of the Positive and Negative Forces
during the Eight Seasons

Eight Cardinal Points Combined with the Eight
Positive and Negative Elements of Matter

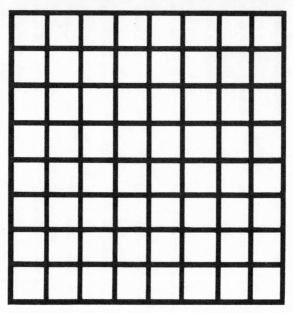

The Unity of the Universe in its Eight Basic
aspects of Time and Space, Force and Matter,
Seasons and Cardinal Points, and their dualities

Final Cosmogonic Pictograph: Division of Universe
into Light and Darkness, Good and Evil

as spring, spring-summer, summer, summer-fall, fall, fall-winter, winter, winter-spring.

The line symbolizing Force, moving downward from the central point, was also multiplied by eight. The division of Force into eight categories introduced a new idea into their cosmogony. They first divided Force into four categories: the energies coming from the stars, those from the sun, those from the earth, and those inherent in man. But each of these four was divided once again, into good and evil. Here appears the first idea of duality, the first dualistic philosophy in the history of human thought, represented by those eight perpendicular lines below the creating point.

When the horizontal line to the right of the point was multiplied by eight, it represented Space divided into the eight cardinal points: east, south-east, south, south-west, west, north-west, north, and north-east.

The horizontal line to the left of the point, symbolizing Matter, became the four elements: air, water, earth, and fire, when multiplied by four. These four were then multiplied into eight, each signifying the good or evil form of the element.

The next movement of the Creator combined the two groups of perpendicular lines, those above and those below the point. Thus the symbols of the eight seasons, combined with the symbols of the eight forces, represented the activity of the positive and negative forces during the eight seasons.

The eight horizontal lines to the left of the point and the eight to the right of it were then made into a pictograph signifying the combination of the eight cardinal points of Space with eight positive and negative elements of Matter. And the final step was to divide these sixty-four squares into good and evil, light and darkness.

This pattern of thirty-two white squares and thirty-two black squares represented to the ancient Sumerians the unity of the universe as it existed around them, in its eight basic aspects of time and space, force and matter, seasons and cardinal points, and their dualities. There could be no simpler way to explain the Creation, yet in its geometric clarity is hidden all the mysteries of the Universe.

According to the ancient Sumerians, there are two aspects—light and darkness—to everything which exists. For instance, light manifests in our bodies as health—darkness as disease. Light manifests in our minds as harmonious thoughts, darkness manifests as disharmonious thoughts. Light manifests in nature as animals useful to man, such as the cow and the horse; while darkness manifests in creatures harmful to man, such as snakes or jaguars.

Everything has two aspects in life, and in our body, our mind, in human society, on our planet, in the universe, everywhere—there is a constant battle raging between the forces of Light and the forces of Darkness. There is a battle between vitality and disease, between our harmonious thoughts and disharmonious thoughts, and so on. At first, this seems a classic dualistic concept, but it is not, really. For darkness is only the absence of light, and disease is only the absence of health. And disharmonious thoughts and emotions are only the absence of harmonious thoughts and emotions, and so on. Therefore, this philosophy not only represents dualism, but also monism.

This is what is called a Cosmic Concept, and I would like you to understand the intricacies of cosmic concepts through the history of different teachings. For instance, I want you to hear a quotation from my book *Toward the Conquest of the Inner Cosmos,* in which I give the quintessence, in ten lines, of the main problems of philosophy:

Essentially there are four solutions to the problem of unity.

(1) There is only spirit: the foundation and essence of everything that exists is spiritual. This is the doctrine known as spiritualism.

(2) There is only matter: spirit itself is merely a manifestation of matter. This is materialism.

(3) Both spirit and matter exist, but outside both is a third element—the divinity—from which they are derived and to which they owe their unity. This is the doctrine of theism.

(4) Both spirit and matter exist, but there is a third element, not outside them, but within them. This is again the divinity—from which they are derived. Matter and spirit are merely the two forms of manifestation of a divinity that has no other existence than through them. This is pantheism.

All these aspects are taken care of in the Sumerian concept of Zarathustra in an extremely interesting way; all the complexities of the greatest mystery of our existence—the origin, the beginning, the Creation—are presented in simple clarity; yet without being simplified.

After the creation of the universe comes the creation of the earth. And here again I would like to say a few words about the difference between knowledge and wisdom. The consensus of scientific knowledge about the origin of life on our planet is that life originated in the ocean, and from the ocean, life propagated to land. In 1933 I wrote a book published in England called *Cosmos, Man and Society,* * in which I submitted another theory about the origin of life on our planet. In that pre-geological era when our planet was a fiery body, surrounded by the cold of cosmic space, the fiery body made impossible any form of life, because on its surface were boiling waters, oceans with unbelievably high pre-geological temperatures. Then it was surrounded by the frigid cosmic space where life was also impossible, as the cold was so extreme it would be difficult to measure even with our present instruments. But between the two extremes, from the upsurging waters of the boiling oceans as they went up from time to time into the cold cosmic space, a vapor-filled atmosphere developed, where finally the temperature reached a level favorable to life. It was not as hot as the fiery body of the planet and the boiling oceans, and not as cold as the surrounding cosmic space. And when the preconditions appeared for a simple, monocellular amoeba, then that simple, one-celled creature made its appearance. Therefore, according to my theory, life did not start in the ocean, but in a pre-geological vapor-filled atmosphere, where between the two extreme temperatures, a zone existed which was favorable to life.

My theory of the origin of life was only one part of *Cosmos, Man and Society,* which had more than 850 pages and dealt with everything from astrophysics and astrochemistry to the social structure of the creative self-sufficient homesteads of the future. Having an incorrigible sense of humor, I will tell you how my old friend George Bernard Shaw put the book to use. He was once accosted by a group of reporters who had heard a rumor that he was taking pills, and since it was known he was a strict vegetarian and never took any medicine, it promised to be a good story. But Bernard Shaw assured them, "Yes, gentlemen, it is true that I was taking pills for my insomnia, I confess it. But no more! Now it is different. All I have to do is open *Cosmos, Man and Society* to the chapters on astrophysics and astrochemistry, and I fall asleep

---

*This large volume has been out-of-print for many years, and is not the same book as the one with the same title available at present from the International Biogenic Society. The present *Cosmos, Man and Society* was also written in the thirties and is one of the most important works of Edmond Bordeaux Szekely.

in ten minutes!" So this is probably the only good use you could have today of that book, because it represents a lot of knowledge, and I think you had better read those books of mine that deal with wisdom, rather than knowledge.

The scientific theory of the origin of life is knowledge. But Zarathustra's interpretation of the origin of life and the origin of our planet, the earth, is *wisdom,* because it gives you what you need to know for your happiness. Every step of the Creation gives you a source of power, a source of harmony, a source of energy, which is at your disposal, and which you can utilize in your life. This is why it is a wisdom teaching—not a dry mathematical, physical, astrophysical or astrochemical treatise—but pure wisdom, which gives you roots, your unity with all the forces surrounding you. According to Zarathustra, we live in a field of forces, and our well-being, our happiness, and our individual evolution depend on how much and how well we absorb the energy, harmony, and knowledge from our surrounding field of forces.

And in the process of the Creation, Ahura Mazda answers the questions of Zarathustra, those questions you heard from the *Zend Avesta:* who created the sun, and the clouds, and the wind. . . all the questions of Zarathustra are answered by Ahura Mazda by following the steps of the Creation. But it is not just knowledge; you are not just told, well, this was created and that and that—but at the same time everything which is created is a source of energy, harmony and knowledge for you. Also, after each step of the Creation there is a suggestion, what you should do with this force and with that power.

So now we have the sixty-four squares, divided between Light and Darkness, representing the Cosmos, the whole Universe. Let us walk down to the garden where you will find a life-size tapestry of the Universe, these same sixty-four light and dark squares, and Ahura Mazda will again create our planet—and the life on our planet. Every step of this Creation will give you a source of energy and harmony, and every step will tell you what to do with this source and that power. It is the Creation of our planet and life on our planet based, not on sterile knowledge, but on living wisdom which you can utilize in your life for individual happiness.

*(The class adjourns to the Tapestry of the Universe, where the Creation of Life on our Planet is re-enacted. Professor speaks the*

*words of Ahura Mazda, and Norma the words of Zarathustra, as*
*one by one, the visible forces of Nature and the invisible forces of*
*the Cosmos are set in their proper places on the Tapestry of*
*the Universe.)*

*Voice of Ahura Mazda:*
I, Ahura Mazda, the CREATOR:
First I have made the Kingdom of Light.

> *Voice of Zarathustra:*
> You shall use your creative powers;
> Your role on this planet
> Is to continue the work of the CREATOR!

The second of the good Kingdoms
Which I, Ahura Mazda, created,
Was Ahura PRESERVER.

> You shall PRESERVE all useful things
> In the Kingdom of Ahura Mazda.
> You shall prevent damage
> To whatever has value, whether a tree,
> Plant, house, love, or harmony in any form!

The third of the good Kingdoms
Which I, Ahura Mazda, created,
Was the Ahura of ETERNAL LIFE.

> To reach ETERNAL LIFE
> You shall have sincerity in all you do
> And with everyone you meet!

The fourth of the good Kingdoms
Which I, Ahura Mazda, created,
Was the Ahura of WISDOM.

> Gain WISDOM
> Through the Good Thoughts
> Of Ahura Mazda!

The fifth of the Good Kingdoms
Which I, Ahura Mazda, created,
Was the Ahura of WORK.

> You shall perform your daily WORK
> With honesty and efficiency!

The Tapestry of the Universe before-the Creation of our Planet, as described in the *Zend Avesta* of Zarathustra. *(below)* Norma places the symbols of the Creator and Preserver on the Tapestry of the Universe, as the Creation begins.

The sixth of the Good Kingdoms
Which I, Ahura Mazda, created,
Was the Ahura of LOVE.

> You shall speak
> Only gentle and kind words
> Through the LOVE of Ahura Mazda!

The seventh of the Good Kingdoms
Which I, Ahura Mazda, created,
Was the Ahura of PEACE.

> You shall maintain PEACE,
> Create it within yourself and around you.
> Prevent inharmony, enmity and violence!

The eighth of the Good Kingdoms
Which I, Ahura Mazda, created,
Was the Ahura of POWER.

> You shall perform good deeds
> Through the POWER
> Of Ahura Mazda!

The ninth of the Good Kingdoms
Which I, Ahura Mazda, created,
Was the Fravashi of SUN.

> You shall expose your body
> To the Golden Rays of the SUN!

The tenth of the Good Kingdoms
Which I, Ahura Mazda, created,
Was the Fravashi of WATER.

> You shall purify yourself with water
> Every morning, and drink every day
> The life-giving WATER of Ahura Mazda!

The eleventh of the Good Kingdoms
Which I, Ahura Mazda, created,
Was the Fravashi of AIR.

> You shall be outdoors
> And breathe the life-giving AIR
> Of Ahura Mazda!

The twelfth of the Good Kingdoms

The last of the symbols are placed on the Tapestry of the Universe, bringing to completion the Creation of our Planet. *(below)* The Creation is complete, with all the Sixteen Forces of Nature and the Cosmos: Sun, Water, Air, Food, Man, Earth, Health, Joy, Power, Love, Wisdom, Preserver, Creator, Eternal Life, Work, and Peace.

Which I, Ahura Mazda, created,
Was the Fravashi of EARTH.

> You shall create more abundant life
> On this EARTH by growing plants
> And begetting children!

The thirteenth of the Good Kingdoms
Which I, Ahura Mazda, created,
Was the Fravashi of FOOD.

> You shall eat living FOOD
> From the gardens of Ahura Mazda!

The fourteenth of the Good Kingdoms
Which I, Ahura Mazda, created,
Was the Fravashi of HEALTH.

> You shall use all good forces:
> Sun, Water, Air, Food, Man, Earth and Joy.
> Harmony with all good forces
> Will give you vibrant HEALTH!

The fifteenth of the Good Kingdoms
Which I, Ahura Mazda, created,
Was the Fravashi of MAN.

> Oh, MAN! You shall strive incessantly
> On the ascending path toward
> The Light of Ahura Mazda!

The sixteenth of the Good Kingdoms
Which I, Ahura Mazda, created,
Was the Fravashi of JOY.

> Be always JOYOUS and happy
> In the Service of the Law!

*(After the re-enactment of the Creation of Life on our planet, the class returns to the outdoor lecture hall, and copies of* The Individual Inventory of Zarathustra, *an excerpt from Professor's book* The Essene Book of Asha, *are distributed.)*

*Professor continues:*
If you look on the next-to-last page of this little booklet, you will find all those pictographs of the Creation which we just saw in process. So please keep it in your hand while we talk about this stage of the Creation.

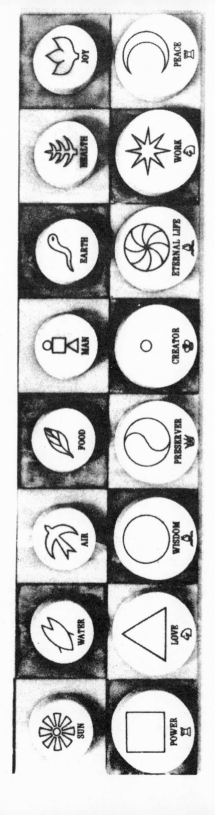

Zarathustra and the Cosmic Order (ASHA)

THE SIXTEEN BIOGENIC EARTHLY AND COSMIC FORCES

(For a complete explanation of these symbols, please read "The Essene Book of Asha," by Edmond Bordeaux Szekely.)

You probably noticed that the simple, visible forces of nature, such as sun, air, water, etc., were mentioned in the *Zend Avesta* as *Fravashis,* which means natural forces. The others representing invisible, spiritual forces, were called *Ahuras.* These are terms taken from the *Zend Avesta.* Although the pictographs actually belong to an age previous to Zarathustra, it was Zarathustra who collected all the existing knowledge and wisdom of that age into the *Zend Avesta,* so this is why we are using terms from the *Zend Avesta.*

That great encyclopedia had a knowledge part, which really represented all the sum total of knowledge of that age—nature, science, astronomy, etc.—and it also had a wisdom part. The wisdom part contained that knowledge which is essential for individual happiness, and these sixteen pictographs are taken from that wisdom part of the *Zend Avesta.* They are the most ancient part of the book because they relate to a previous age in which alphabets and writing did not exist, only pictographs. In my translation of the *Zend Avesta* of Zarathustra, this part is given in pictographs at the end of the book. You will see that every line of the English translation is marked with a number, and that number is marked also in its corresponding pictographic section, so anyone can read the English text and the pictographic text simultaneously.

What is interesting is that this part of the *Zend Avesta* relating to wisdom, was written with these sixteen pictographs, and anyone reading the two texts together will see the superiority of the pictographic language over our contemporary language. All the complexities, all the philosophical content of the *Zend Avesta* is written by the variation, permutation, and combination of these sixteen pictographs.

Each of these pictographs, as you saw, represents one step of the creation of our planet and life on our planet. At the same time, each of these pictographs represents one source of energy at our disposal, for us to use in our lives—one force in the field of forces surrounding man. Each one carries with it some advice for man—how to use it, what to do with it. And at the same time, each one represents an idea, a building brick, the combination of which, in their sixteen-fold form, make up the wisdom part of the *Zend Avesta* of Zarathustra—all the knowledge you should have for your individual happiness and individual evolution.

If you look at these pictographs you will notice that the figure of man is composed of a circle, a square, and a triangle. If you remember, the square represents Power, the triangle Love, and the circle Wisdom. Also, you will notice that Man is before the Creator, right before the Creator. According to Zarathustra, the role of man on our planet is to continue the work of the Creator. Because Creation is perpetual—the creation of life, all creation is constant and perpetual. It is both the duty of man and his privilege to continue the work of the Creation on our planet. At the same time it represents the idea that right behind man is all the power of the Creator as a source of energy for man—that man at any time can ask the help of the Creator—he is right behind you.

You also probably noticed that there are affinities between each cosmic force and the natural force in front of it. For example, here we have Power—and before Power is the Sun. On our planet the best manifestation of power is the sun.

In front of the symbol of Love is Water. Water has affinity with Love. Remember in the Old Testament—love is stronger than the currents of deep waters—love is stronger than death. The ancient traditions of Sumeria, which through the Babylonian Prison were transmitted to the people of Israel, established always that affinity between love and water.

Before the symbol of Wisdom is Air. The ancient Mazdean traditions maintained that the precondition of wisdom is the right breathing. When we breathe in the right way, it sends oxygen to our brain, and that is a precondition of our brain functioning properly. So there is always an affinity between wisdom and air.

Then there is an affinity between the Creator and Man. Man was created by the Creator and is similar to Him, because man has the most important job on this planet: to continue the work of the Creator, to continue on our planet the work of creation. It is to Man this great work is given, because it is in Man that Nature became conscious. This is the title of nobility of Man. Of all the species, only Man reached that degree where nature became conscious. Man is able to understand nature, to understand the laws of nature, and is able to utilize the forces of nature. It is Man who created something above existing nature, which was created by the Creator, Ahura Mazda. Man is also the creation of Ahura Mazda, but Man created something plus above nature, which we call Culture—that is the creation of Man. Man is the only species on earth who was able to create something plus, above nature. This is why every person born today inherits tremendous wealth. Every one of us is a billionaire, because we inherit the great masterpieces of universal culture—we inherit the works of Socrates, Plato, Aristotle; we inherit the works of the *Vedas,* the *Upanishads,* the *Tao Te Ching;* we inherit the knowledge of the *Zend Avesta,* the *Essene Gospel of Peace,* the works of Michelangelo, Leonardo da Vinci, Rafael, Rubens; the music of Bach, Mozart, Beethoven; we inherit the plays of Shakespeare, the novels of Victor Hugo, Tolstoi, Dostoievski—we inherit a tremendous wealth of universal culture. And this universal culture was not created by Ahura Mazda, but by Man who continues the work of creation, the work of Ahura Mazda on our planet. This is why Man stands directly in front of the Creator.

The work of the Creator, as well as our task of continuing the work of the Creation, would all be futile if there were no Preserver, a very important force on our planet. If we would just go on creating things and then forgetting about them, they would all perish. It is the Preserver which preserves all good things on our planet, all the great things created. And the visible expression of the Preserver, preserving life, is Food. Food is represented by a grain of wheat, just as air is represented by a flying bird, and water is represented by a fish. All these pictographs are real, are living— they represent something which exists around us, not just abstract letters: a, b, c. Food is preserving the life of man, the life of the birds, of the fish, the life of everything on our planet—and this is why it stands before the Preserver.

Before the symbol of Eternal Life is Earth—because life on earth has an affinity with cosmic universal life. Universal life, life in the Cosmos, life on earth—share a common bond and origin. We shall not suffer in the megalomania that life exists only on our planet, because we have billions of planets in our galactic system, and we have billions of galactic systems in our universe, and to say that only on this single insignificant point in the infinite cosmic space there is life, and nowhere else does life exist, would not only be megalomania, but illogical and ignorant, as well. There is a Cosmic Ocean of Life, according to the *Zend Avesta,* which means a kind of solidarity between all forms of life on those billions of planets which have life in the universe. And there is a solidarity between all forms of life on our planet. That beautiful cypress tree near our gate is alive—it has the same vitality we have—

*Saint Francis of Assisi*

the same vitality another living being has on some faraway planet. In that Cosmic Ocean of Life there is a great affinity and solidarity between all living beings. This is why Earth stands before Eternal Life, to symbolize the affinity between universal life and life on earth.

The next pictograph is Work, and before it, Health. The affinity between these two is extremely important, because Work is at once a terrestrial function and a cosmic function. Look at a plant: there is a great deal of work going on—the solar light falling on the leaves, the process of photosynthesis, then the little plant draws water from the rain, draws minerals from the earth, and there is a continuous work of growth, a tremendous laboratory of life going on in that little plant.

The same thing goes on in our organism. The solar rays fall on the oily surface of the skin, which creates a biochemical substance called ergosterol, which is carried inward into the metabolism and transformed into Vitamin D. Life forces are continuously working in ourselves, in a plant, in a tree, all over our planet. And it is Work which is maintaining all living beings, the same Work which is the foundation of our health. It is our most important precondition.

Finally, we have the symbol of Peace. And before the symbol of Peace we see Joy manifested. Peace is the most important quintessence of the teaching of Zarathustra, of the *Zend Avesta,* of the Essenes. You remember the Essene Gospel of Peace. Even their greeting was, "Peace be with you." This is our most noble role. We must achieve inner peace. We cannot continue the work of the Creator if we do not have inner peace. And there is nothing more joyous than Peace. Peace manifests in Joy. Joy is Peace manifested. Joy—that overflows in the Chorale of the Ninth Symphony of Beethoven, to the words of Schiller's Ode to Joy—the same joy that sings in the words of St. Francis' beautiful Hymn to the Sun, when, in exuberant joy he gives thanks to the Creator for Brother Sun, Sister Rain, Brother Wind, and all the beautiful things for which we have reason to be joyous.

In fact, in view we are almost to the end of the session, I want you to hear this Hymn to the Sun of St. Francis, and each time you read it I want you to remember that Joy is the greatest way to praise the Creator—Joy is the greatest revitalizing power.

The Hymn to the Sun was written in beautiful 13th century medieval Italian, but I have translated it into English for you:

Praise be to Thee, O Lord,
for all Thy creatures,
And especially for our Brother, the sun,
who gives us the day,
and who shows forth Thy light.
Fair is he and radiant with great splendor:
To us he is the symbol of Thee, O Lord.

Praise be to Thee, O Lord,
for our sister, the moon,
and for the stars.
Thou hast set them clear,
beautiful and precious in the heaven above.

Praise be to Thee, O Lord,
for our brother, the wind,
for the air and the clouds,
for the clear sky and for all weathers,
by which Thou givest life
and the means of life
to all Thy creatures.

Praise be to Thee, O Lord,
for our brother, fire,
by whom Thou givest us
light in the darkness.
He is beautiful and bright,
courageous and strong.

Praise be to Thee, O Lord,
for our sister, water,
who is so useful to us,
humble, precious, and chaste.

Praise be to Thee, O Lord,
for our mother, the earth,
who sustains and nourishes us,
bringing forth divers fruits,
flowers of many colors,
and the grass.

This exquisite song of praise from the greatest of the saints
shows his intuition, because all those brothers and sisters he

addresses are, of course, the Angels of the Essenes—the Angel of Sun, the Angel of Water, the Angel of Air, and so on. He never knew about the Essenes, nor did he ever read the Essene Gospel of Peace. But two thousand years later, with his profound intuition, with the power of evidence in an outburst of joy, he thanks the Creator for all these beautiful powers which are also described in the *Zend Avesta* of Zarathustra (which he never read, either). It is a good example of one of the three yardsticks to truth: intuition. Great geniuses all have this kind of intuition. This is one of the eternal masterpieces of universal literature, unique in its purity and simplicity, and so much in affinity with the Essene Gospel of Peace and the *Zend Avesta* of Zarathustra.

I want to say a few more words about this last symbol of Peace, which was always the center of the Essene teachings. It was in the Babylonian prison that the people of Israel absorbed the teachings of Zarathustra and the *Zend Avesta,* as Babylonia is in the same geographical area as Sumeria, where these traditions lingered for many thousands of years. And of all the Ahuras of the Kingdom of Light of Zarathustra, they felt the most affinity for the Ahura of Peace. You remember in my book, *From Enoch to the Dead Sea Scrolls,* there is a chapter on the Sevenfold Peace. This Sevenfold Peace was extremely important for the Essenes, because they lived in an era of wars and violence, described by the Roman historian Livius as *bella omnium contra omnia*—everybody's war against everybody. And so they accentuated Peace and utilized its symbols in many ways: the Tree of Life with seven branches and seven roots, the seven lights on the sevenfold candlestick, and their traditional greeting, "Peace be with you." And we must not forget the Gospel of Peace. It was not without reason that they were called "the people of Peace." Their tradition is not anti-quated, nor obsolete. We need it today a thousand times more than they did, because for the first time in history mankind has acquired the means, the capacity, to destroy our planet. With the menace of thermonuclear war on the horizon, "everybody's war against everybody" takes on a new and deadly meaning. This part of the Creation, the meaning of the Ahura of Peace of the *Zend Avesta* of Zarathustra in our daily lives in the twentieth century is of paramount importance.

I want to add one more thing before we conclude today's discussion. Maybe you don't know about it, but here in Costa

Rica preparations are being made to establish a World University of Peace. It is not accidental that this idea materialized in Costa Rica, because it is the only country in the world which has no army nor munitions factories. Peace is a way of life here and has been for centuries. The idea has been approved by the United Nations, the land exists for its disposal, and concentrated efforts are being made here in Costa Rica to bring to fruition as soon as possible the first World University of Peace. You see, it is not enough to be a pacifist and talk about peace. We have to be active in our pursuit of peace: we have to do something about it. A philosopher said that the main trouble with the world is that the bad people are always more active than the good people. This is why the idea of the World University of Peace is so appealing to me, a peaceful philosopher. It will teach the practical aspect of Peace, as well as all the philosophical ones. Practical knowledge is what each one of us needs in order to become an active point in the universe and increase the Kingdom of Light.

It is an idea, which like all great ideas, excels in simplicity. If we want to save mankind, we must train people in the practice of peace, not only in its theory. We have to utilize all the possible sources we have, in literature, in philosophy, in religions, in the arts, to use all these which inspire peace, instilling them on a large international scale.

Our most important problem at the present for mankind is peace—because everything else will be in vain, everything in the world which is accomplished in any field, if suddenly somebody pushes the button and we go up in smoke and *adios amigos,* no more mankind and no more life on earth. This makes extremely important one part of the Essene teachings, which we discussed in previous Seminars and probably you read about in the books: the Sevenfold Peace.

The Sevenfold Peace is the most complete and universal program of establishing peace. It starts with *Peace with the Body,* because if we have no peace in the body we are sick, diseases are fighting for our vitality, and we are unable to do much for peace. The second is *Peace with the Mind.* If peace doesn't exist in our mind, in the form of harmonious thoughts and emotions, then again we cannot do much for peace outside. *Peace with the Family,* with our friends, or we may say, with our immediate environment, is also very important, because as Romain Rolland so beautifully

*May 22, 1979: Professor and Norma visit the President of Costa Rica, Lic. Rodrigo Carazo Odio, to discuss the establishment in Costa Rica of the World University of Peace.*

said, the whole depends on the position of the atoms composing it. And unless we can establish peace around us, with our friends, with those we come in contact with every day, unless we can achieve this, there is not much hope for mankind to have peace on a large scale. Then *Peace with Humanity*—it means a kind of normal, logical system of society like that of the Essenes, who had no slaves, nor lords, nor poverty nor riches—a system in which the economic conditions of peace can exist. Here come the teachings of the Essenes concerning the simple, creative way of living—which I developed in my books as the creative health homestead, published in England under the title *Cottage Economy,* in the U.S. as *Father, Give Us Another Chance,* and most recently, *The Greatness in the Smallness.* Then we must also have *Peace with Culture,* which is very important. We must draw from the great values of universal masterpieces and put them into practice, trying to avoid the mass of cacophonic, disharmonious, mediocre flow of literature which is causing a lot of trouble, disorganization, confusion and dullness in the human mind. All this I point out in my book, *Books, Our Eternal Companions.* Then, *Peace with the Earthly Mother* is vital—because if we continue to pollute the air, the water, the soil, the foods, and everything around us, in that case it doesn't matter what kind of perfect social system we may have a hundred years from today, we will all be perishing sicklings on our planet and will not survive. And last, *Peace with the Kingdom of the Heavenly Father*—the seventh peace of the Essenes which is so important—to establish contact with the Cosmic Ocean of Life, with the Cosmic Ocean of Thought, with the Cosmic Ocean of Love—to absorb the greatest teachings deriving from the Cosmic Order, as Zarathustra so beautifully said. This is the reason why we united these things in one of the smallest countries in the world, where peace is a way of life—where there is no army nor manufacture of arms, and where there will be one day the World University of Peace.

## The Third Day: July 28, 1979

Well, we will continue where we left off yesterday. If you remember, we discussed the Creation of the Universe, starting with the Creator creating Time, Space, Matter, and Force, and the eventual division of these basic elements into the Tapestry of the Universe. Then we went outdoors to the Tapestry of the Universe and watched the Creation of the Earth and Life on Earth unfold. Today we will discuss the Creation of Man.

If you remember, according to the *Zend Avesta,* Ahura Mazda created the sixteen forces, a totality of a field of forces surrounding us. And if you look at the pictographic chart, you will see that Man, one of the sixteen forces, is directly in front of the Creator. This means that we have a duty and a right at the same time: a duty to establish harmony with all these sixteen forces, and the privilege to absorb these sixteen forces as sources of harmony, energy, and knowledge. According to the *Zend Avesta,* we live in a dynamic universe, and we shall know how to adapt ourselves to each of these forces which surround us and flow toward us permanently. According to modern psychology, intelligence is the ability to adapt ourselves adequately to unexpected changes in our environment. And this parallelogram of forces which surrounds us is constantly changing. The intensity of the solar rays are changing, the temperature of the air we breathe is changing, and so on. Therefore, the first thing we have to know and understand is each one of these sixteen forces. Then, once we understand them, we have to become receptive to them. If we want to utilize them as sources of energy and harmony and knowledge, we have to feel these forces as they penetrate our bodies and our consciousness. I will give you an example. The Ninth Symphony of Beethoven, or his Missa Solemnis, is a tremendous spiritual experience. It is an inner experience equivalent to the revelation of the Sermon on the Mount, provided we are receptive to it. If we are not receptive to it, it is just a lot of sounds which mean nothing to us. In the same way, although we are surrounded throughout our lives by the sixteen forces, unless we learn to be receptive to them, we cannot utilize their energy, harmony, and knowledge.

First, we have to realize that these sixteen forces are realities. Realities. There is no doubt that we breathe air, that the sun shines on us, that it rains and we have water, that we eat foods, and so on—it is evident that Zarathustra does not want you to believe in or be receptive to imaginary or unreal things, but only to the realities, to the tangible forces which surround us constantly.

Once we are receptive, then we must simply find the most practical ways to utilize these forces. And to accomplish this, there are three attitudes we must have: we must understand them, we must feel each force and become receptive to it, and we must finally learn the most practical way of applying the force.

The first attitude is to understand each force, so it becomes a source of knowledge to us. For example, we shall study the air—learn the quality of air in different places—realize that we are breathing 12,000 quarts of air a day, and that it is our most important food. We shall realize what effect pure air has on us, and then what effect polluted air has on us, etc. We shall realize the effect of water—we should know that a hot bath has a relaxing effect, while a cold bath has an invigorating effect. If you look through my book, *Healing Waters,* you will see there about eighty different applications of water, and that each has a different effect on the different parts of the human body. And also we shall understand the sun—how to take a sunbath, how not to take a sunbath, and so on. We shall realize that the sun is the source of all the energies on our planet. The solar rays falling on the leaves of plants creates an interaction, a process we call photosynthesis, which fixes different energies from the atmosphere, creating plant substances. The plants absorb air and sunshine and rainwater, minerals from the soil, and due to the process of photosynthesis they will grow with great speed. Without photosynthesis we could not have vegetable life on this planet, nor animal or human life. We indirectly eat sunshine when we eat green foods, green leaves, green vegetables—all these are indirect sunshine. But the rays of the sun also have a direct effect on our bodies, when the solar rays fall on the oily surface of the skin. That creates a biochemical substance called ergosterol, which is carried inward into the system and transformed into Vitamin D. And Vitamin D has control over the metabolism of calcium, and indirectly over the metabolism of phosphorus, and presides over a great number of basic biological functions. Well, I could go into a great many more details, but I just mention these things as examples that we must know and understand all these forces. If we don't understand them, we will not be receptive to them, and we will not be able to apply them.

*(At this point, Norma draws Professor's attention to a quart jar full of sprouted whole wheat grains which she intends to grind and bake into loaves of "Zarathustra bread," so named because its description was found in the Zend Avesta. She passes the jar around so everyone may see the one and only ingredient of the bread she will make.)*

Now I will stop for a moment to go a little bit on a side track. The reason why we always associate Zarathustra with wheat is that it was he who, through many variations, created the perfect form of the wheat grain and brought it to its optimal form. If you read my book, *The Essene Book of Asha,* you will find there a beautiful legend, the *Legend of the Wheat of Zarathustra.* Wheat became the staff of life in all ancient cultures. At the present it is no more the staff of life, of course, because bread today contains all kinds of chemical humectants, anti-oxidants, emulsifiers, preservatives, etc. But in ancient Persia, and in all ancient civilizations, bread, real bread made only from whole grains, was the main food, the staff of life. We have in the *Zend Avesta* a very interesting recipe—how to make the so-called Zarathustra Bread, as it was known all through antiquity. I did a special study, because I have always enjoyed reconstructing foods eaten by people in ancient civilizations. For instance, I reconstructed the *hydromel,* that drink made from honey, which in ancient Greece they gave to the athletes before going to the Olympic games. In the same way I reconstructed the bread of Zarathustra, used in ancient Sumeria and ancient Persia. I remember once at Rancho La Puerta I was preparing these different foods eaten in ancient civilizations, using archeological techniques, and my old friend Aldous Huxley came by to visit with a friend of his. He looked at what I was doing, tasted my hydromel, tasted my Zarathustra Bread, and then he said to his friend, "Do you know what Professor is practicing here? It is a new branch of medical nutrition: gastroarcheology!" Now I thought it would be a good idea, as we talk about Zarathustra and study his teachings, that you get acquainted with the bread of Zarathustra, from a recipe several thousands of years old. The recipe may be ancient, but Norma will go now to the kitchen, grind these sprouted wheat grains which you have been looking at, and toward the end of today's session we can all taste a freshly-baked sample of my gastroarcheology! The bread of Zarathustra is made from nothing but wheat—but germinated wheat, or as it is called today, sprouted wheat, The process of sprouting mobilizes all the dormant life forces in the wheat, and it becomes alive with the maximum potency of enzymes, hormones and vitamins. It is nothing else but sprouted wheat, but you may not believe it when you see the texture and discover how good it tastes. I want you to realize that a food such as bread is also an archeological object. What people ate in ancient civilizations is as important an object of archeology as where they lived, in what kind of houses, or what baths they took, etc.

Now once we understand a force, then we can become receptive to it. When we understand the sun, then when we take a sunbath, we will know exactly what is going on in our organism, and we will feel all the benefits of the sunbath. Just as you must know the musical notes to feel the spiritual experience of the Ninth Symphony of Beethoven, the same way with a natural force, like the sun. An example of receptivity and the importance of feeling each force is faith healing. It is a reality. Faith can heal. But why, and how? When someone is receptive to a force, even if that force has no outer reality, the very belief in the force can affect healing in the organism. Here, incidentally, is the tremendous difference between the teaching of Zarathustra, based on sixteen real forces, on realities, and other teachings based on hypotheses. We cannot dispute about the sixteen forces, for they are realities which have the power of evidence and physical experience. Nevertheless, when someone believes that something can heal his organism, no matter what that something is—even if it is a match-box—then an interesting thing happens: the organism starts to throw into the circulation and metabolism different enzymic hormonal substances, and these enzymic hormonal substances have a regulating effect on our metabolism and in our biochemistry. Faith healing is nothing mysterious or mystic. We know exactly what is happening in the human body. Now if belief in non-real things can very frequently create these beneficial enzymic and hormonal effects, and result in certain cases of healing, provided the preconditions exist, how much more these sixteen forces of Zarathustra, which are realities—sun, air, water, food, etc.—can make us receptive if we understand them. And being receptive creates a deep feeling so we know exactly what these forces are doing in our organism. Zarathustra was the greatest realist of all the great teachers of history. He advised us to understand and to feel and to believe only in real forces—not theological dogma or philosophical theories.

The third thing we must learn about the sixteen forces is how to apply them in a practical way, in an optimal way. Unless we learn how to actually, practically, optimally apply the forces in our daily lives they cannot benefit us, no matter how much we may understand them or feel receptive to them. For an example, let's go back to that sunbath. Fine. Practical experience will show you that you will receive a better effect if your head is toward the north, and your feet toward the south—not for any mystical reason, just simply because in this way the angle of the solar rays reach your body optimally, and this very sensitive group of nerves, the cerebral system, etc., which is protected by the cranium, will not be exposed to the hottest sun, and you will also feel much better. Then you will find out, also by experience, that you have a better effect if you take a sunbath, we say, five minutes on the front, five minutes on the back, five minutes on the right side, and five minutes on the left side. Much better than if you stay the whole twenty minutes on one side. And of course I already mentioned the beneficial effect of the sun on the tonus of the skin, the pigmentation, the texture, and most of all, the circulation. When the circulation is improved, then the blood will distribute and carry everywhere with greater efficiency the hormones, enzymes, vitamins and minerals throughout the body. Therefore, sunbathing has a multiple beneficial effect. But at the same time, if you don't understand or have the knowledge of the nature of the sun, if you don't know how to use it, you can have several disasters. First of all, if you have a disposition to high blood pressure, the sun will raise your blood pressure to dangerous levels. Number two, you may easily get a sunburn. Number three, you can even develop skin cancer by staying too long in the sun, or exposing your body indiscriminately to solar rays which are too intense, for instance, at the wrong time of the day. Or you may take a sunbath in the wrong frame of mind. If your mind is receptive, and you understand the physiological effects of the sun, then the sun will relax you. But if, while you are taking a sunbath, you are thinking, oh, I'd better go to the bank in an hour and transact some business and sell some stock because the stock market is going down—then instead of relaxing, you will develop nervous tension, and the sun will only make worse the situation.

Therefore, you can see that all these forces, each of these

sixteen realities, can be extremely beneficial, a source of energy, a source of harmony, a source of knowledge to us. And they can also be a source of disaster. In this dualistic universe there is darkness as well as light, and every one of these sixteen forces, which continuously surround us with their forcefields, can be beneficial, or harmful, depending on how they are used. In the *Zend Avesta* of Zarathustra, he tells us exactly how to use these sixteen forces, and how not to use them. This is an extremely important thing to realize.

Now I want you to remember the sixteen acts of the Creation, which we re-enacted on the Tapestry of the Universe. Whenever Ahura Mazda created one of these sixteen forces, at the same time he gave a statement of advice, a dominant motif, how useful the force is, what we shall do with it, and so forth. Of course, in the *Zend Avesta* it goes into great detail, but if you remember, at each step I quoted from the *Zend Avesta* that I, Ahura Mazda, have created this and that kingdom, and the name of the force. Then Norma put the pictographic symbol representing that force on the respective place on the Universe—the Tapestry of sixty-four white and black squares representing Light and Darkness. And at the same time she read a quotation from the *Zend Avesta*—what is our privilege and what is the good deriving from the respective act of the Creation. The concept of Zarathustra is an all-sided concept, an omnilateral concept—it starts from the beginning with the Creation of the Universe, the Creation of Earth and Life, and continues with the Creation of Man, and what Man is doing to continue the work of the Creator. Zarathustra expects us to *create ourselves,* to work constantly on our self-improvement and self-perfection. The same forces created in the beginning, at the beginning of the beginnings, these tools which Zarathustra gave us, form the most important system for the creation of man into ever-higher degrees of individual evolution. And we are performing this creation and re-creation of ourselves at every moment of our existence, because whenever we do something, we can do it in the right way, in harmony with the Kingdom of Light, or in the wrong way, in harmony with the Kingdom of Darkness. When we breathe, we may breathe pure air, with all the tremendous benefits, and we may breathe polluted air. We may eat the right food, with all the tremendous benefits, and we may eat the wrong food. We can practice love in our lives, and we can practice hatred. We can practice wisdom, and we can practice ignorance. Everything we do may belong to the Kingdom of Light or the Kingdom of Darkness. We may do something which may improve our health, and we may do something which may create disease and deteriorate us. As we are moving and navigating in this forcefield, the sixteen forces are given to us as our inheritance—but how to use them, *if* we use them, or if we use them in the wrong way, that is entirely our own act. That is man's free will.

So to help us cooperate always with the Kingdom of Light, Zarathustra created a system of "individual inventory," of self-analysis, which is the most tremendous tool for our self-perfection. I say tremendous, because it is all-sided and all-encompassing—it embraces everything, all aspects of life. It was practiced many thousands of years ago by the followers of Zarathustra. Once a week they made an individual inventory—they glanced through their activities of the week and reflected on how they had used the sixteen forces—what they used correctly, what they used incorrectly—and through this all-sided, optimal method they were able to greatly further their individual evolution. I cannot overstate the importance of the all-sidedness of the individual inventory.* Even Socrates, one of the greatest philosophers of all time (and one of my old friends), although he uttered the classic command, *gnothi seauton* (in ancient Greek, *know thyself*), he never told us how to go about it. He also said that an unexamined life is not worth living—that is a wonderful statement—but he didn't tell us how to examine ourselves and in what way we shall live. But Zarathustra did. Zarathustra gave us a concrete, precise program for living, accentuating the fact that it is up to us whether or not we create harmony with these sixteen forces. He emphasized that everyone gets results in the measure of his individual efforts. No one is perfect. I don't think there is anyone who established one hundred per cent perfect harmony with every one of these sixteen forces. Perfection is an ideal. We may never reach it, but we can go closer and closer to it, every day. And this is what Zarathustra wants us to do. He wants us to use this arsenal of sixteen forces as sources of knowledge and harmony and energy, so we can go closer and closer to perfection. Now it is up to us—if we want to

---

*A reprint of the Individual Inventory of Zarathustra from *The Essene Book of Asha* may be found in the Appendix of this book.

have good health, if we want to create our happiness, if we want to have a meaningful life and absorb all the sources of energy and harmony and knowledge—or if we are satisfied to do just half a job or a fraction of it—if we want to harvest the benefits of all the things we did well, or suffer the consequences of all the things we did wrong.

From one aspect, the individual inventory is the most ancient form of psychoanalysis, seven thousand and some hundred years before Freud. Now Freud reduces everything to one factor: sexual repression. You know very well his theories, so I won't go into details—repression of sexual impulses creates neuroses, etc. etc. That's fine, I don't doubt he was a good psychiatrist, and he is right in many ways, but his approach is one-sided, while the system of Zarathustra includes everything, all the factors, all the aspects of life. A German writer and philosopher, Lessing, said, "In this world everything has so many angles that we just cannot understand how they fit together." Well, that seems to me the attitude of modern man—he gave up long time ago trying to fit together all the things of the world, and is just floating, like a fallen leaf in the wind. And whenever he meets some kind of philosophy or theory he believes in it until he finds another kind and switches to that one, until he meets still another idea and falls in with that one. But in the general disorientation and confusion he just cannot figure out how all these things fit together. Well, Zarathustra did figure it out. Zarathustra, with his all-sided, omnilateral teaching, does fit together all the loose ends.

But the teaching of Zarathustra is purely empirical—it is like a laboratory work—you can experience it yourself as you do these things Zarathustra advised, and immediately have the benefits, or immediately have the consequences, the wrong ones. There is nothing more important than the empirical method, the practical experience. Remember what the great poet, Goethe, said: "Gray are all the theories, but green is the tree of life." Never in any other age did I find so many complex, complicated, confused books on psychology and philosophy than in this last half of the twentieth century. Visiting the marketplace of books published these days, I can only paraphrase the words of Diogenes when he visited the marketplace of Athens—I never knew there were so many books I don't need! The greatness of the teaching of Zarathustra is that you can practice it, you can do it, you can reap the benefits—and if you do it wrongly, you suffer the consequences, and that too is a lesson for the future. And you have the arsenal of the sixteen forces to correct all the wrong things you did before. This is the greatness of the system of Zarathustra—the all-sidedness, the omnilaterality—because the greatest mistake we can make in life is one-sidedness.

Now let us have a little glimpse of these sixteen forces. I will let Norma discuss this with you, as she always did it in other Seminars, a little less heavier than I do.

*Note from Norma:* As the tape recorder was left on only when Dr. Szekely was speaking, my contribution regarding the individual inventory was not transcribed. But looking back over our manuscripts, I found something written several years ago—an adaptation of some material by Dr. Szekely about Zarathustra, the *Zend Avesta,* and the individual inventory—material included in *The Art of Asha: Journey to the Cosmic Ocean,* which I helped to edit. Written from the standpoint of a pilgrim on the eternal path of individual evolution, I think it will give a clear picture of just what it means to do the individual inventory on a regular weekly basis. In reading about each force, remember that there are always three questions one must ask about each, three questions which pertain to the thinking body, the feeling body, and the acting body: (1) Do I understand the power or force? (2) Do I feel the significance of the force deeply and sincerely? Am I receptive to it? (3) Do I use the force continually and in the best possible way?

## SUN   *the reflection of Power*

I try every day to take a short sun bath, knowing that the sun improves the tonus of my skin, creates Vitamin D in my organism, and lulls my nerves into beautiful relaxation. I feel the life-giving energy of the sun enter my body. I am aware that to use this force in the optimal way it is important to have just the right amount of sun: it can be my best friend or my worst enemy. But when I use it correctly, my body, mind and soul open in joyful thanksgiving for the gift of life.

## WATER   *the reflection of Love*

In the morning, as soon as possible after awakening, I bathe in clear, fresh water, following one of the primal instincts of Nature for cleanliness. In the splash of a shower, the warmth of a bath, or the refreshment of a dip in an icy stream, water invigorates, cleanses, purifies. I understand the necessity of water to the earth; how barren deserts are turned into beautiful forests only through water, and I have rejoiced to see little birds splashing happily in rain puddles. This same thirst of Nature is present in my body, and I try always to use water—in eating, drinking and bathing—in its freshest, purest form.

## AIR  *the reflection of Wisdom*

There was a time when I did not think of what kind of air I breathed—only that I did breathe. It did not disturb me that in smoke-filled rooms, in streets full of car exhaust and on smoggy days, my lungs, for protection, took in only the tiniest amount of air necessary for existence. Now that I know the tremendous importance of deep breathing of pure, fresh air to every one of my physical, mental and spiritual functions, I try at every opportunity to surround myself with pure air, Nature's most important food. Truly it is said, "Where there is breath, there is life; where there is life, there is breath." I am breathing 10,000 quarts of air daily, and my lungs will only be used to their natural, full capacity if the air I am breathing is free of impurities and smoke. Knowing that my mind and body will not develop without the nourishment of pure air, I try to be just as careful of the air I breathe as I am about my food being free of dirt.

## FOOD  *the reflection of the Preserver*

The cells of my body are made of what I eat; therefore, if I wish to be strong and healthy, I must make sure my body is built from the building blocks of Nature—wholesome, natural foods: proteins without animal fats; sprouted whole grains and seeds; raw vegetables and tender baby greens; fruits, as they are from trees and vines; nuts, honey and dried fruits for energy. Foods which have been refined, chemically processed and overcooked are empty foods, and though they may please an artificially developed taste, they can only deplete my body, prematurely age it, and rob it of its natural resistance. Therefore, I try my best to give my body only those foods which are as close as possible to their original, natural state.

## MAN  *the reflection of the Creator*

I know that the primary role of my life is to continue the work of the Creator on earth. I know that in some unique way I have something original to contribute to the good of society, be it in music, science, art, literature or education. No matter if my talents are large or small, unusual or prosaic, I must try to discover my hidden potential and use it actively to creatively enrich the world in which I live. Only in untiring efforts to further my progress in life can I truly find fulfillment as a human being, and I know that no one can do the job for me. Sometimes the realization that only the individual efforts of man, including myself, can change the world for the better, is frightening and awesome; still I know that this most glorious challenge of all is within my power, and my diligent efforts to continue the work of creation will not go unrewarded.

## EARTH  *the reflection of Eternal Life*

Reproduction, in its many forms, is one of Nature's greatest miracles, and I must learn to respect it, learn from it, and live in harmony with it. In vegetation, especially the wonder of grass, which blankets the earth with the magic of growth, I will learn many secrets of Nature, and I will try to keep a small garden for this purpose. Even if it is just a window box, this attempt at growing things teaches me about many aspects of reproduction. In my body, I know that sexual energy may be used harmoniously in two ways: first, to raise a family in accordance with the laws of power, love and wisdom; and second, to use one's own sexual energy to regenerate the body and find an outlet in the creation of beautiful works of art, literature or music. In other words, whether I decide to use sexual energy for reproduction or regeneration, the important thing is that I use it wisely and harmoniously.

## HEALTH  *the reflection of Creative Work*

I want more than anything else to be healthy, for I know that only if my body is healthy can I be an active point in the Universe, working to further my own evolution and the evolution of society. Perfect health consists of cooperation with all the natural forces: sun, water, air, food, man, earth and joy; also good health rests in balance and moderation: I know that I need exercise to maintain good circulation; I also need plenty of sleep to "recharge my batteries" and renew my energies. A good diet of thoughts and emotions aid my digestion of foods, so it is equally important for my health that I cooperate with the cosmic forces as well.

## JOY *the reflection of Peace*

Although it is sometimes difficult, I try very hard to be joyous at all times, for I know how great is my power to influence those around me. When I am sad, it does no good to pretend—I sadden those around me, turning their happy moods into bad ones, affecting their nervous systems and even their digestions. When I am happy, I bring as if into a magic circle those around me into my happiness, watching them go away with new courage and strength. When I have negative feelings, I try to think of them as a contagious disease, with which I do not want to infect others. And I know, deep within, that charity, sermons and good deeds mean very little unless I can give to my fellow man the gift of my own happiness. And what do I need to become happy? Simply the understanding and awareness that life is a precious gift, a privilege that is not to be wasted; and as Marcus Aurelius said, it is not the things around us that make us happy or unhappy, but our attitudes toward those things. Therefore, I shall continue trying to radiate joy wherever I go, for the world desperately needs it as never before.

## POWER *the mirror of Sun*

In the Bible it is written, "By their fruits ye shall know them," and I know that unless the private universe of my thoughts and emotions give birth to good deeds and noble actions, my purpose as co-worker with the Creator is not being fulfilled. Stendahl said, "One can acquire everything in solitude—except character." In the language of Zarathustra, solitude is the silent realm where thought and feeling flow together in their own rhythm of music and fire. But all the beauty of mind and soul are useless unless they are translated into action—action to eliminate human suffering, action to work for the betterment of the human race and the planet on which we live. The present chaos and constant wars can never be controlled by peaceful attitudes alone; action is needed, *my* action, not someone else's. I know that I do not need to travel across the world to work for good and be an active point in the universe—my family, my friends, my place of work, all provide me with a vast arena where I may slay innumerable dragons of ignorance, injustice and disharmony and, through example, awaken in others eternal sources of energy, harmony and knowledge. Zarathustra writes in the Zend Avesta, "Evil exists not, only the past; the past is past, the present is a moment, the future is all." Only by my good deeds and actions in the present can I mold the future of which Zarathustra spoke.

## LOVE  *the mirror of Water*

Just as the force of Power is manifested in deeds, so is the force of Love expressed in words. I know that until I have attained psychological maturity, it will be impossible to love every person I meet in life with wisdom and detachment, but by learning to allow only loving words to express my thoughts, I can actively cooperate with the force of Love, and better my own life and the lives of those I come in contact with. Love is unity; in the Cosmic Ocean of Love all forms of life are united—life itself is an expression of love. Therefore I remind myself each day that it is not important whether I agree with, disagree with, approve of, or disapprove of my fellow man. The main thing is that we are all cells of one body, the Cosmic Ocean, and a growing awareness of this basic oneness will more than anything else develop my ability to love. I try to follow Buddha, who advised that each time before speaking we ask ourselves three questions: "Is it true, is it needful, is it kind?"

## WISDOM  *the mirror of Air*

To always have good thoughts is the essence of Wisdom, but oh, how much more difficult it is than to say good words and do good deeds! For words and deeds may be controlled by discipline, but thoughts have a way of entering unbidden and departing without leave, and I realize that until I can rein in the wild horses of my thoughts, I am no better than a slave in a harshly ruled country. So I try and try, and keep trying. Knowing that two things cannot exist in the same place at the same time, my strongest weapon is to strengthen the good in all I encounter, instead of trying to fight the evil. By trying to let only harmonious thoughts enter my mind, I am refusing to give reality to negative, destructive thoughts. Thoughts are immensely powerful currents, and it is up to me to create heaven or hell for myself and others, depending on the quality of my thoughts. When temptation arrives in the form of depression, despair or sadness, I try to "tune in" to the thoughts of the great masters, philosophers, musicians, artists and poets of history, feeling their ageless power rescue me from the mediocre and banal thoughts of all those who live lives of "ignorance in action." In the *Zend Avesta* it is written, "In what fashion is manifest Thy Law? O Thou Great Creator! By good thought in perfect unity with reason, O Zarathustra!" By always seeking out the wise, the good, and the beautiful, I will gradually learn to conquer my thoughts—for until I do it is of no use to conquer anything else.

## THE PRESERVER  *the mirror of Food*

In my efforts to understand and cooperate with the law of preservation, I realize more and more why the Preserver on the Ashaic chessboard has more power and versatility than any other figure. All that has been created, everything of beauty, everything of value must be steadily guarded and maintained, or the act of creation has no significance. The exquisite ceiling of the Sistine Chapel of Michelangelo took years to create and profoundly affects the spiritual lives of millions, yet a careless bomb in a needless war could destroy it in seconds. A friendship is not a static thing but a dynamic, changing relationship, and a reckless word may damage it beyond repair. Friendship, love, works of art, whatever the creation, the word has no meaning without preservation. Sometimes I am helpless to prevent waste and destruction, as in the case of war, yet in my own sphere of activity I can use my influence in countless ways to protect, preserve and practice prevention; grass, trees, flowers, a house, a relationship, whatever it may be, material or immaterial, the act of Preservation is the indispensable twin of Creation. Every time I cooperate with the Preserver, the words of the *Zend Avesta* take on new meaning: "The Keepers and Preservers of the Earth shall restore the World! When Life and Immortality will come and Creation will grow deathless!"

## THE CREATOR  *the mirror of Man*

I am told that my role on this planet is to continue the work of the Creator. Immense and awesome and nearly impossible does this role seem to me, for I am not a Beethoven, a Shakespeare, nor a Leonardo da Vinci. Nevertheless, by utilizing whatever small talents I may have and struggling to uncover my hidden potential, I am unconsciously setting in motion my creative powers. If my desire is great and my efforts sincere, no one can predict in what areas and in what paths my creative ability may develop. Again the *Zend Avesta:* "The Creative Mind within us is causing the Imperishable Kingdom to progress." From these words I know that the Creative Mind is slumbering within, waiting to be awakened through incessant striving. As long as the goal of my efforts is the Imperishable Kingdom, the realm of Creation may well be mine to rule, explore, and obey.

## ETERNAL LIFE  *the mirror of Earth*

This is the only force that assails me with unanswerable questions: What exactly does Eternal Life mean? Is it the perpetuation of my species or my own personal eternity? If I am to live forever, why do I not remember having lived before? And so on, and on. Though my immature need of security will continue to ask, the answer of the *Zend Avesta* is simple: "Let the enlightened alone speak to the enlightened." In other words, metaphysical speculations about Eternal Life are irrelevant; to cooperate with this force means that I must live my life eternally, here and now, as if each day were my last, savoring every precious moment, utilizing each minute of every day to help eliminate human suffering and teach others through example. It was Goethe who said, "To reach the Infinite, one must tread the Finite in every direction." To understand Eternal Life, I must first understand everything about this life; instead of asking futile questions about what happens when I die, I will make my own eternity in the present, by accepting myself as I am without rationalization, by approaching all my tasks with sincerity and honesty, and not shirking my responsibilities. "...and the same inexorable price must still be paid for the same great purchase..." The price of Eternal Life is sincerity, honesty and incessant striving in this one.

## WORK  *the mirror of Health*

Those who consider work as a hardship and a burden will never be able to understand this force. "Only a select few can rise to the conception of work as the supreme realization of the mind." How the numbers of that "select few" would swell if all people could experience the joy and radiance of creative work! "Happy is he who has found his task; he should not ask for any other blessing." Only through work am I able to put into practice the principles of power, love, wisdom, peace and joy; only in work can I move among men and teach the things I have learned. The man who complains about the nature of his work complains without reason; if the will and desire are strong enough, any kind of work will reflect the Law, and even the lofty roles of Teacher and Priest are worth only as much as they are in harmony with the Law. When I feel gratitude and love surge within me for the gift of life and the pursuit of knowledge, I know that only work done well can prove the sincerity of my feelings. The *Avesta:* "What is Thy Kingdom, O Creator? What are Thy riches? Thine Own, in my Work, in my Holy Service."

## PEACE  *the mirror of Joy*

I know that deep, lasting peace within myself can come only when I live in true harmony with all the sixteen natural and cosmic forces. It is the result, the crown, the golden aura that surrounds the mountain of hard work and striving. Still, as I tread the path of daily effort, there will be many opportunities for me to either contribute to the cause of peace or to that of discord; a simple word, a gesture, a calm judgment may bring communication where there was a lack of it, settle an argument that would have swelled and multiplied, or bring harmony into an atmosphere usually devoid of it. People who constantly live on the sharp edge of nervous tension and anxiety are slowly being poisoned by their emotions. My peaceful efforts to soothe, calm and relax are balm more important than pills or potions. My own Peace, solemn and mysterious, may only be at present a whisper of the future; yet I can give what I have—peace to answer the chaos around me, peace to still the frantic clamor of those who seek security in material things, peace to strengthen the spirit of those who desperately want to find their Home but know not how. I feel very certain that though I am but a pilgrim and I make many mistakes, the very act of trying to understand these sixteen forces, of feeling their power in my life, of making my actions manifest their purposes, gives me the power to guide, direct and teach in a way I never would have thought possible. This is my peace—even before reaching my goal, of being on the Path—and knowing that to travel this Path hopefully and joyously is my glorious mission and the mission of all mankind.

*Norma's description of the individual inventory comes to an end and Professor's lecture resumes, after a question asking about the reverence felt by Zarathustra for trees, and their unquestioned position of importance in the philosophy of the Zend Avesta.*

Zarathustra used the tree as an example. The tree is the Law itself. Why? Because the tree absorbs all the sources of energy in the universe—absorbs the solar rays, the air, the minerals from the soil, the water from the rain; and the tree, enriching the soil every year through its fallen leaves, returns everything he receives to the ground, and much more. The tree at the same time is not only uniting in himself all the forces of nature, but also brings us fruits and gives shade. The tree is also an expression of peace. I don't think you ever saw two trees fighting with each other. The tree is the Law itself, an example for us how to live in peace and absorb all the sources of energy around us. This is the tree.

When I first came to this country, the First Lady of Costa Rica asked me, could you write for the children of Costa Rica a book to instill in them the love for trees, so when they grow up they will protect the trees and we will have a better chance for our ecology? Well, I told her I had written all kinds of books about archeology and cosmology and cosmogony, but I never wrote a book for children—nevertheless, it was a fine challenge, so I told her I would try. So I wrote the book, called *Hermano Arbol,* or *Brother Tree.* I gave her as a gift a thousand copies and she distributed them in a great number of schools, leaving a copy of it in hundreds of schools everywhere. Then my publisher told me, well, we also have children in the United States—why don't you translate your book from Spanish to English? So I did it, and it appeared under the title *Brother Tree.* It was illustrated by a very talented painter who painted frescos in many cathedrals in Mexico, and although this book required something on a slightly less imposing scale, she did a really excellent job. Children need pictures, drawings— they have a better understanding that way.

*Est modus in rebus*—there is a way in all things. The philosophy of Zarathustra is all-sided, omnilateral—not multilateral— omnilateral. Through history it was applied in many ways, for many purposes. And everywhere it worked and everywhere created great masterpieces. We will soon show you an example of one of these applications, but first I want you to understand something. You remember Zarathustra said that the noblest of all professions

Do not cry, my children.
Love is stronger than death.
Life always starts again!

—from *Brother Tree*, by Edmond Bordeaux Szekely.

is to be the gardener of the earth. Gardening was the center of the practical teaching of Zarathustra. According to Zarathustra, a garden is a microcosmos. It has everything from universal life. The gardener is working with all the forces of nature—is working with the soil, is working with the sun, with the water, with the air. He must take into consideration a parallelogram of forces of different intensity, different frequency, of sun, water, air, and soil. When we gather up a handful of topsoil, we have there also a microcosmos. In a minor scale we may have in our hand whole galactic systems, whole universes—for that microscopic micro-organism, the handful of earth may be like a galaxy, and the interaction of these billions of microorganisms is what creates the biogenic sphere, the life-generating parallelogram of forces. In a handful of soil there is not only the whole planet in minor scale, but it also teaches us something: through the vegetables and fruits we grow and eat, the particles of our organism are being created by the particles of planet earth. Planet earth was created by the particles of the sun, and the sun was created by the particles of galactic substances of centripetal ultra-galactic substances. Therefore, we have a unity of the macrocosm, the ultra-galaxies, the sun, the earth, and the human body. We are one. We are the same.

Now, when we are gardening, we continue the work of the Creation, because we are creating life, living plants. We are utilizing these cosmic substances which did reach us after billions and billions of years through the ultra-galactic system, the galactic system, the sun, the earth—and we are an active point in the universe—we continue the work of Creation—and the garden is a unit of life. Zarathustra demonstrated all the laws of the universe and life through the garden. The greatest poet of modern India, Rabindranath Tagore, created a university of peace, Santiniketan, where students still study under huge banyan trees, and that is very beautiful, but Zarathustra went farther—Zarathustra used the garden to demonstrate, like in a laboratory, all the laws of life, utilizing all the forces of terrestrial nature. Zarathustra was a master gardener, not only because he described in the *Zend Avesta* a complete treatise of natural gardening, but because most of all he was a genius of simplicity—he was able to teach the most complex laws of life and the universe through his garden. It was there that his disciples went to study with him, and it was there where he demonstrated daily the most essential laws of life and the universe.

Gardening was a kind of religion with the ancient Zarathustrians. It became an art, and was always a very meaningful activity. You probably know a little about Japanese landscape gardening, and the use of different shapes of rocks, plants, and water to create different effects—it is a faraway echo of the landscape gardening of ancient Persia. But the gardening of ancient Persia was a cosmic function; it symbolized the unity of man and the cosmos, and it was the means of translating knowledge, of teaching, not only how to grow in the optimal way fruits and vegetables for eating, but even more important, what the function of gardening does to man, the gardener, while he is gardening. It taught him understanding of the meaning of life and the universe, and his dynamic unity with the forces of the universe and the whole cosmos.

The products of this religion were hundreds and hundreds of fruits and vegetables which we know today. I studied the origin and migration of every fruit and vegetable through history—it is a fascinating study—and I found to my greatest surprise that the majority of our fruits and vegetables which we know today came through migration from ancient Persia, of Zarathustrian origin.

Now I would like to describe to you this aspect of the ancient Zarathustrian religion and philosophy, gardening, which was practiced for thousands of years in ancient Persia. It is tremendously meaningful, teaching much more than you would think possible. We will use this white table and small squares representing the sixteen forces, because according to an ancient Chinese proverb, a picture says more than a thousand words.

The garden is represented by the white table, and the plants, trees, flowers of the garden by the sixteen symbols, because the ancient Persians had a representative plant, flower or tree for each of these basic symbols, which they considered the building bricks of the universe, and the building bricks of life on our planet. For instance, sun was represented in the garden by a sunflower, which is turning continuously toward the sun. Water was represented by watercress, which lives in the water, and so on. There is an interesting sentence in the *Zend Avesta* which says that the apple is the king of the fruits, and the grape is the queen of the fruits. So to represent the Creator and the Preserver in the garden, they used the apple tree and the grapevine. I don't want to go into details, but each one of the natural and cosmic forces was represented by some member of the vegetable kingdom. To create a garden was a work of art. To create a garden was a sacrifice to Ahura Mazda. To create a garden was thanksgiving to Ahura Mazda for all these beautiful forces he created for us. To create a garden was to create a microcosmos. Gardens of today can be beautiful, but they are not meaningful, they don't express a philosophy of life.

Now you will see demonstrated one of my methods of teaching archeology through miniatures. We will make a journey back to ancient Persia, and all together we will create an ancient Zarathustrian garden.

*Professor has on his desk two containers, one with little wooden blocks, each with the name of one of the sixteen forces, the other with letters and numbers representing each of the sixty-four black and white squares of the Tapestry of the Universe (see diagram), although for aesthetic reasons the table used is pure white. To create the garden, each of the seminar participants takes at random one block from each container, placing the cosmic or natural force on the respective place on the table-garden.*

This is to demonstrate that in whatever field we apply the methods of Zarathustra and the *Zend Avesta,* with innumerable variations, combinations and permutations of the basic forces and symbols, we could go on for billions of years without repetition. Now pick up one from here, and one from here. Every square is represented topographically by a letter and a number—you can see on the side it is marked with numbers from one to eight, and here by letters from A to H. *(as Norma suddenly leaves for the kitchen to avoid burning the bread)* I will suspend the Cosmic Order while you turn the oven off!

*(As each participant takes one wooden block from each container, Professor reads aloud the name of the force and its position on the table, and the photograph shows the finished Zarathustrian Garden created by this method.)*

Each of you planted a plant, or a tree. This is one of the many billions of possible varieties of a Zarathustrian garden of ancient Persia. Try to imagine that each of these symbols represents a real shrub or plant or tree or vegetable, and then this has profound philosophical and spiritual meaning. Everything which is in the garden represents a cosmic and natural force. It is a microcosmos.

Professor explains how Zarathustra demonstrated all the laws of the universe and life through the gardens of ancient Persia. *(below)* In miniature, the tabletop "garden" expresses a beautiful and meaningful philosophy of life.

They used these meaningful gardens for meditation. Thousands of years passed, and ancient traditions usually have a tendency to become pale and sometimes deformed. Nevertheless, to a small degree these things survived. There is a large Zarathustrian community in Bombay in India, and when I walked on the streets there I immediately knew that in this house lived a Parsee, a follower of Zarathustra, or that one there, because there was always a garden behind the house. Of course, they forgot a lot of traditions. For instance, the present-day Zarathustrians always have a fire burning in their homes, to worship Ahura Mazda. But when Zarathustra said that we must always feed the Fire of Life with pure wood, he didn't mean to pick up some special wood and keep an actual fire burning—he meant that we have to keep up our vitality with pure foods. This was the deep meaning of Zarathustra, but of course through thousands and thousands of years it became pale, though some automatic habits did survive. It is the same with the garden. I could see that there a Parsee lived, a follower of Zarathustra, because they always had a beautiful garden beside the house, and everything was always kept extremely clean, while the other houses were often like slums, absolutely dead. The Parsees were the most active. Even with those traditions very pale and deformed as they survived, they still represented an excellence in their social environment.

Now I want you to use your imagination and see, instead of this table with its symbols, a beautiful garden full of plants, trees, flowers and shrubs, all representing the sixteen natural and cosmic forces of Zarathustra. The ancient Persians used the garden every morning and evening, walking around in it to have communion with the different plants, communions representing these forces, and through the garden they refreshed themselves spiritually and physically. Naturally, they also benefited by eating the fruits and vegetables grown in the garden, but most of all the garden was a sanctuary, a temple, a place of meditation, of relaxation, a place to absorb different forms of energy from all these forces of the *Zend Avesta.* It was also a reminder to establish harmony with all the forces which the plants or trees represented. It was a stimulus to carry on with the individual inventory. If you remember, all these pictographs on the table represent the different forces of the Creation which we took part in yesterday. At the same time, they represent the different forces with which you shall create harmony, sources of energy and knowledge and strength in your individual inventory, the Zarathustrian system of psychology and self-analysis, and program of living. And the same forces are represented in the Zarathustrian garden. This is what I mean when I say the teaching of Zarathustra is omnilateral, all-sided.

*(Norma appears with two loaves of freshly-baked Zarathustra bread, the same bread which appeared earlier in the session in the form of sprouted whole wheat grains. She passes the loaves around, and each delighted participant breaks off a piece to eat.)*

Now the only ingredient of this bread is wheat. You see, you go to buy a loaf of bread these days and read the label, you have a whole national museum there—all kinds of things which would inspire Diogenes to say, I never knew there were so many things I don't need. And not only is wheat the only ingredient, but every part of the wheat is here. The wheat germ is not taken away, the bran is not taken away—it is a complete wholeness—a whole food. And it is so easy to make.* Usually, people use flour to make bread. But even if it is whole wheat flour, it is not fresh—you will find it difficult to wait with your bowl at the mill to take it home as soon as it is ground—and when whole wheat flour is packaged and kept, then air particles enter into the flour and create oxidation of the wheat germ oil. And all polyunsaturated oils, including wheat germ oil, become rancid and unhealthy when they oxidate. You have no guarantee, even if it is of high quality, that the flour you buy to make your bread is fresh. The only way to be sure of freshness is to grind the sprouted wheat yourself, just before you bake the bread, as Norma has done today. This bread is made directly from the wheat grain. The wheat grain has the endosperm. the mesosperm and the exosperm, and it is surrounded by the bran which protects the wheat germ from oxidating—it is a good example of the wisdom of nature. Therefore, when you use the whole wheat grain, and sprout that grain, you immediately mobilize the life forces of the grain, and there is no time for anything to oxidate. And you don't need any yeast because the process of fermentation developed during the germination creates its own yeast. In the whole field of gardening, plants, vegetables, fruits, grains—this was the greatest masterpiece of Zarathustra—

---

*Recipes for Zarathustra Bread and its simpler variation, Essene Flat Bread, may be found in *The Book of Living Foods, The Chemistry of Youth (Search for the Ageless, Volume III)*, and *Biogenic Reducing*, all available from I.B.S. Internacional.

the wheat. This is why the force of Food is represented in the Cosmic Order by the wheat grain. This is why the legend of the wheat was so important in the teaching of Zarathustra. And now you can taste what we made of it. All these facets of the teaching of Zarathustra were important thousands of years ago, but they are much more important for us today; in this one-sided world we need the omnilaterality, the spirit of all-sidedness of Zarathustra.

*Why does the bread taste so sweet, when it contains nothing but wheat?*

It is true—it is hard to convince people that it doesn't contain sugar. The sweet taste comes from the process of dextrinization of the starch, the transformation of starch into sugar, which also makes digestion easier. When the wheat is sprouted just a little longer, it can taste more like cake than bread. But there is still nothing in it but wheat. Of course, you can use your imagination when you bake the bread—there are unlimited combinations. You can add raisins, figs, dates, all kinds of healthy things. There is a Latin proverb: *de gustibus non est disputandum*—never dispute about taste. What one person likes added to the bread, someone else may not like at all, so you have to use your creativity and improve your relations with the Creator on the field of Asha.

*How long will wheat grains stay fresh in storage?*

The wheat berry itself, before it is sprouted, can last a long time, depending on the climate where it is stored. In a dry climate, wheat grains really last for a long, long time. You probably heard about archeological excavations where wheat grains were found which were still able to germinate after thousands of years— remember, the author of the wheat was Zarathustra! We never had a greater gardener than he was. Next to him, maybe the second greatest gardener in history, was Luther Burbank. He created hundreds of different plants and vegetables and fruits, and of some fruits a great variety—of the same fruit he created dozens of varieties, all with different characteristics. He wrote six volumes of his experiences as a gardener, and really, from the standpoint of gardening, I may say he was the Zarathustra of the twentieth century.

Now we will go back to our ancient Zarathustrian garden, and its relationship with the individual inventory. Here again comes *de gustibus non est disputandum*—because some of Zarathustra's disciples did very well with their individual inventory in some fields, but were not so good in others. Therefore, in their gardens they planted those things which represented their weak points, so when they went to meditate every morning and evening in the garden, they could constantly have before them that weak point which was growing before them. There is a beautiful part of the *Zend Avesta* which tells you, when you see a tree representing one of the forces, to go to that tree and put your arms around it, embracing it. So those who needed inspiration in certain fields planted those trees or plants which represented those areas—it was certainly not an ordinary garden, but a very meaningful one. Each garden was individual and was intended to be a source of advancement of individual evolution. There was a unique and intimate relationship between a Zarathustrian and a tree, plant or flower in his garden. Everything in the garden is magic. You don't have to look for magical squares, because there is no greater magic than life in action. Every unit, every plant in the garden is the creation of nature, with elements as old as millions of years. For instance, the ordinary grass arrived to our planet fifty million years ago—and without it, no life would be possible—the ordinary grass. Other plants have other ages, but there is no greater miracle and no finer magic than the evolution of life on our planet from that first little monocellular amoeba—the progress of a tiny microscopic being to a Michelangelo, or an Einstein, or a Beethoven. Now the same thing is happening in the vegetable kingdom. You go back and you will find that it took hundreds of thousands of years of evolution to create the journey from an insignificant seed to a tremendous walnut tree. That is true magic. There is no greater magic than that of the kingdom of nature.

I want to read to you those few beautiful lines from the *Zend Avesta* describing that intimate relationship between man and tree. It is from the Vendidad, Fargad II:

> Ahura Mazda made the answer:
> Go, O Zarathustra,
> Toward the high-growing trees,
> And before one of them
> Which is beautiful,
> High-growing and mighty,
> Say you these words:
> Hail be unto thee,

O good living tree,
Made by Mazda!

The tree was a sacred symbol, a sacred plant, a sacred power of Zarathustra. According to the *Zend Avesta*, just as in the animal kingdom man represents the highest form of creation, so in the vegetable kingdom the tree is the most perfect species. Therefore, according to the *Zend Avesta*, man should live with trees, and eat their fruits, which are the most perfect symbiosis between man and tree. In one of my books, *Cosmos, Man and Society*, there is a chapter, *The Symbiosis of Men and Trees*, in which I translate the ideas of Zarathustra.

Speaking of that chapter, I remember in 1936 I gave a summer seminar in England, Sir Stafford Cripps, who was Chancellor of the Exchequer at that time, founded an International Health and Education Centre—he invited the best, the most brilliant students from all over the British Empire, to teach and train them, and send them back so they could carry on teaching in their own countries. At the time I was living in the south of France and Sir Stafford invited me to give a seminar every summer at his Centre, which was in Leatherhead, Surrey. Well, at the end of one of my lectures, a gentleman approached me and said, "You know, I just read three days ago that chapter in *Cosmos, Man and Society* about the symbiosis between men and trees, and I want to dedicate my life to the protection and propagation of trees on our planet." That was Dr. Richard St. Barbe Baker. He founded a worldwide movement called *Men of the Trees*, and he is really an active point in the universe. He was fighting with the lumber barons in northern California in order to preserve the redwoods—those beautiful trees as old as the age of Abraham—and he intervened with the government and was able to save them. Years ago, he arrived in Iran with little flats of thousands of little tree seedlings, and after thoroughly confusing the customs officials by declaring their value as greater than all the riches in the world, brandished a letter from the Shah which invited him as an honored guest to personally supervise the forestry project in Iran. His most grandiose plan which he is still working on is the reforestation of the Sahara desert. He shuttles back and forth between New Zealand, Africa, England, the United States, and once a year he always sees me and brings me up to date about his adventures. He also used to come to Rancho La Puerta and all my Mexican workers loved him—they called him "El Abuelito," little grandfather. So you see, I wrote that chapter in the spirit of Zarathustra, and it was a fine seed that fell on fertile ground. Now he is in his nineties and still running around, fighting for conservation all over the world—protecting trees and planting trees—and when you meet him, he shakes your hand and says, "Good morning! Let's plant a tree!" I never thought when I wrote that chapter I would have such a wonderful disciple who has done so much good in reforestation of our planet.

Now we arrive to a new application of the teaching of Zarathustra. After the Creation of the Universe, the Creation of Earth and Earthly Life, after the Individual Inventory, after Gardening, we arrive to Art.

The Greek genius, among many other things, created art—art upon which is based all art in our western culture. The artists of the Renaissance borrowed from classic Greece and from there it finally reached us. Art reached a pinnacle in ancient Greece— beautiful statues, architecture, painting—you may remember the story about Apelles who painted a grapevine that was so perfect, the birds came trying to eat the painted grapes.

Now let us go back to ancient Egypt, long before Greece—you will see if you look at the works of art of ancient Egypt that everything is static and rigid. The tendency of form-giving predominates. At the other extreme is modern, contemporary art— there, the dominant tendency is to expression, and when the modern artist wants to express something, he explodes the form. So we may say that the rigidity of exclusive form-giving and the explosive effect of the expressionist are the two extremes. And the classic Greek sculpture and painting, and that of the classic Renaissance based on Greek art, represent a happy balance between the two. We find in these perfect forms, but also tremendous expression. There in western art we did reach the zenith, a harmonious synthesis of the form-giving which predominated in ancient Egypt and the tendency of expression exploding in contemporary art. The whole history of art is a kind of balance or imbalance of the tendency to form-giving and the tendency to expression.

But we have neglected for a long time, centuries and centuries, another source of art—not the Greeks—but ancient Persia. There we find the Zarathustrian traditions of art. According to the *Zend Avesta,* art must not only have expression and form, but it must also be meaningful.

To explain this better, I want you to think about the two main schools of modern art. We have there the cubist school and the expressionist school. The cubists are trying to be meaningful by distilling reality; by reducing reality to the basic geometrical forms of the universe, they go down to the meaning of what they want to create. According to the cubist, all the things that exist, the multiplicity in the world and the universe, all are irrelevant—the only things which are essential are the basic geometrical forms. That is the cosmos for them. So when you look at a cubist painting you will see squares, circles, triangles, cubes—the basic distillation eliminating all that is irrelevant, all the multiplicity of things, and going down to the geometrical root of existence.

Now the expressionist school is completely different. The expressionist has an inner spiritual experience, and he wants to express that inner experience, and is trying to convey to the spectator that experience that takes place within. The form is not important—only the effort to give substance to an experience that takes place in another realm.

You may be thinking about Chagall, Kandinsky, Picasso—and all the other modern artists—they are meaningful, in their way, because they reached a tremendous conclusion—both the cubist and the expressionist—that art is not only the imitation of the material world—art is not only for decorative effect—but art is to express a world conception. Art stands alongside philosophy and religion as an equivalent way to express our outlook on life and the universe.

Of course, the most interesting thing is that once again Zarathustra got there first! Thousands of years ago he was using art side by side with philosophy and religion as an expression of the Cosmic Order. Now I guarantee you that Kandinsky or Chagall or Picasso knew nothing about Zarathustra, nor did they ever read the *Zend Avesta*. They are original people and are not imitating Zarathustra. Nevertheless, chronologically they are about seven thousand years late. Remember when I mentioned to you the "Big Bang" theory of the Creation, and that after centuries and centuries of research the scientists finally reached the top of the Himalayas and found there Zarathustra smiling and welcoming them and asking why it took so long for them to find it? Well, we can tell the same allegory about modern art. Modern artists finally found the real purpose of art—because art is as important as religion or philosophy. Avoiding

Aristotelian logic and syllogism, art can express a concept on the universe and life. The cubist does it in his way, the expressionist in his way.

But Zarathustra, as I mentioned, was omnilateral, all-sided. He utilized the principles of both, and he used art as he used religion or philosophy. In Zarathustrian art they used the basic building bricks of the universe, just like the cubist—the square representing Power, the triangle representing Love, the circle representing Wisdom, and so on. All these things are here. But we have also ideograms, which represent not only the building bricks of the universe, but are also expressing forces. Why? Because the universe is force and matter, a combination of the two, and both of them are contained here. Zarathustrian art has unlimited variations and combinations.

*(Again the white table is used, this time to create a Zarathustrian painting. Each of the participants chooses at random from each of the two containers the name of a natural or cosmic force, and the letter and number indicating its position on the table. The photo shows the finished painting, a particularly beautiful and meaningful one.)*

Using the cosmic building bricks and symbols of Creation, different steps and forces of nature in different variations bring out different meanings in the painting. Now a cubist will be very happy because he will find there the square, the circle and the triangle—his tools. But so will the expressionist be happy, because the paintings of ancient Persia were purposeful—they were not hung on the wall just as a decoration—but were used in the practice of meditation—meditation on the forces of the universe. The ancient Persians always had a room which was used only for meditation—and in that room was always a beautiful, meaningful painting, symbolizing different aspects of life and the universe. In fact, even today the Parsees keep a special room in their houses where they have the burning fire of Ahura Mazda and always some painting on the wall, although they lost the original meaning of both a long time ago. The present day Parsees are not archeologists and they have only fragments of the ancient scriptures, but they do as well as their understanding leads them. All religions suffer from a similar deviation from the original teaching; our present day Christianity is far away from the first century Christians, and the present Buddhists are far away from the original teaching of

Buddha. For instance, in Tibet they use prayer wheels, forgetting the original meaning of Buddha's symbol of the Wheel of Life—they completely forgot the living wheel of natural and cosmic forces and think that by moving a mechanical wheel they will advance their individual evolution. So everything with time is getting dusty and deficient and deformed; nevertheless, sometimes you can still find meaningful paintings in Bombay in certain homes. I had at one time a high priest of the Fire Temple in Bombay visiting me—Dastur Bode—and he brought me pictures of paintings he had on his wall, as well as pictures of his fire altar.

At that time I was working on the completion of one of my books, *The World Picture of Zarathustra,** and he was so astonished to find the original meaning of what he was doing in the Fire Temple, and the real meaning of the Fire of Zarathustra, that he stayed with me for ten months to be able to read the proofs as they came back from the printer.

Now let us look at this painting we have all created together. It can be construed as a cubist or an expressionist painting, because it has pictures from nature—it has the bird, the fish, the tree, the flower—and it has also ideograms—the circle, the square; the triangle—but it is much more than just the obvious cubist and expressionist symbols. It represents the infinite variations, combinations and permutations of the forces—the building bricks of the universe. It is art which is not only beautiful, but meaningful—and the variations are unlimited. It is not just decorative, but serves a profoundly important purpose when it is hung on the wall and used for meditation. It takes us back to the root of all things. And it also represents a powerful field of forces—a force-field which changes every time you make a new painting. Here is the tendency of form-giving, but also the tendency of expression; and the main thing is this, with which all the modern schools of art agree: that art should be the expression of a world-conception alongside philosophy and religion—an equivalent way of expressing a concept of life and the universe. And through this very brief excursion through the ancient Persian world of art, I have tried to give you glimpses of the richness of the Zarathustrian concept of life and the universe.

*The original edition is no longer available, but the entire text of *The World Picture of Zarathustra* may be found in three current books: *The Essene Book of Asha, The Zend Avesta of Zarathustra,* and *Archeosophy, a New Science.*

An ancient depiction of Zarathustra and the Fire of Life, from Professor's *The Essene Book of Asha. (below)* The finished tabletop with the symbols of an ancient Zarathustrian painting, a particularly beautiful and meaningful one.

Music, also, played a very important role in the expression of the Zarathustrian world picture. Each of the cosmic and natural forces was represented by a musical accord—each very brief. So, by using the same methods, just as we made a painting, we can create symphonies, again in unlimited variations, and each tremendously meaningful. Music has a special meaning, because music is the least material art. A sculptor needs stone, a painter needs the canvas and brush and colors, a writer needs pen and paper. But music is immaterial—you cannot see it, you cannot touch it—it is an inner spiritual experience without the mediation of the senses. This is yet another example of the omnilaterality of Zarathustra.

Now I must mention one more thing which developed from Zarathustrian traditions: it is less important than gardening, less important than art and music, less important than the individual inventory, but nevertheless very interesting. I speak of the game of chess, the most ancient game. Those who play it today, even if they are great champions, I guarantee you they have no idea of what they are doing. Because through thousands of years that game has undergone a great many changes, but it was originally borrowed from Zarathustra. I would like to show you the original form of chess in all its ancient purity and simplicity.

*(The picture shows the original chess figures placed on the board, the white figures representing the natural and cosmic forces of the Kingdom of Light; the dark figures representing their negative counterparts in the Kingdom of Darkness.)*

*(after a question about the dark figures)* Darkness is only the absence of light—I will explain the meaning of the dark figures in a moment. Now this was the most ancient form of chess in ancient Persia. Forget the names of the pictographs written there and look only at the figures. You will remember that these are the original pictographs, the different steps of the Creation of Ahura Mazda. From ancient Persian literature, Firdusi, who wrote the *Shah Nama,* the national epic of Persia, describes in beautiful poetic language this most ancient form of chess (you can find it in my book, *The Essene Book of Asha*). And when Omar Khayyam read it, being a philosopher, he had the basic intuition to understand the hidden meaning of chess, and that it is not only a game—in the eleventh century he wrote these beautiful words:

107

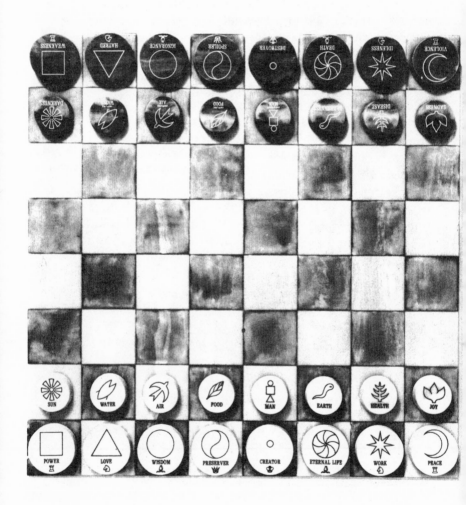

We are no other than a moving row
Of magic shadow-shapes that come and go
Round with the sun-illumined lantern held
In midnight by the Master of the show.
And we are pieces of the game he plays
Upon this chequer-board of nights and days;
Hither and thither moves, and checks, and wins,
And then another Cosmic Game begins.
Ormuzd in wondrous strength and brilliance grows,
Whilst Ahriman in deeps of darkness goes,
And he who guided all with his own hand,
He knows about it all—he knows—he knows!

You can find in ancient Persian literature exactly the different steps of the evolution of the chess game. With time it became a war game, because it was the game of kings, and their favorite occupation was always war. Of course, poor old Zarathustra would turn in his grave if he knew what happened to his cosmic order—that it would be used to represent war games.

I want you to realize one important thing: I studied four thousand years of Persian literature looking for references about the evolution of the chess game, because I knew exactly from where it came—not only Firdusi and Omar Khayyam, but many others. I said that the chess game is the least important application of the Zarathustrian system; nevertheless, even the least important application is tremendous, because here is the Cosmic Order—the sixty-four squares of light and darkness, the universe, created by the Big Bang of time, space, force, and matter—and its octuplication—everything becoming eight, according to the ancient Persian mathematical system—then every step of the Creation is here—the sixteen kingdoms created by Ahura Mazda. And each kingdom represents a field of forces. Suppose we put a chess figure, say, this symbol representing Love, somewhere on the board—if you know the moves of chess then as you move the figure you are symbolizing the action of love, the forcefield of love, moving according to the will of man, upholding Light and opposing Darkness, traveling through Life and the Universe. All the moves of the figures are fields of forces. Each building brick of the universe is here, and each has a living field of force which it creates as it moves.

A chess game is nothing else but the battle of opposing forces

Professor tells the fascinating story of the *Art of Asha,* used in ancient Persia and Sumeria to teach the laws of life and the universe. Each time the forces of Light and Darkness meet in battle, cosmic dramas, symphonies, and works of art are created, continually furthering and enhancing our individual evolution.

of light and darkness, and even after thousands of years of transformation, it is still a dynamic concept. However, the original idea of Zarathustra, the concept which is overlooked today, is that darkness is only the absence of light—it has no power of its own. If you remember, I didn't deal with the dark forces until now—but let us look at them. As I said, darkness is only the absence of light, weakness is the absence of power, hatred is the absence of love, and ignorance is the absence of wisdom. Here we have work—and we look for its counterpart and see that idleness is the absence of work. And here is one appropriate for the twentieth century: violence. Violence is the absence of peace. So really, these dark figures do not have reality—they are simply absences of the real forces of the universe. Nevertheless, they have reality inasmuch as we give reality to them. For instance, if you don't cooperate with love, then you create hatred. If you don't establish harmony with wisdom, you create ignorance. It means that we create all our enemies ourselves, which is a superfluous luxury. The greatness of Zarathustra was to point out that we have to cooperate with the sixteen superior forces, the forces of Light—and in that case, the forces of Darkness will not exist. But when we do our individual inventory and we see that we have gotten upset and we hate somebody, well then, we did not cooperate with love—we have created a monster: hatred. If we neglect wisdom, then we become ignorant, and ignorance belongs to the Kingdom of Darkness, and so on. So although chess is the latest historical phase and application of the method of Zarathustra to teach the laws of life and the universe, it still has the great values of the original concepts of Zarathustra.

Here is an interesting experiment you can try, and a good way to illustrate the deep meaning of the original form of chess: there are books containing historical chess games played by great chess masters of the past. Substitute for the modern names, such as King, Queen, Bishop, Knight, and so on—the original names—Creator, Preserver, Eternal Life, Love, etc. And then when you reconstruct the same chess game you will have instead a cosmic drama—an enthralling battle between the different forces of Light and forces of darkness—it will be something like Faust, a wonderful allegory of man's struggle to overcome gravity and reach the Cosmic Ocean. Any chess game played can be transformed into a cosmic drama in this way. And you should also realize that since the dawn of history there have not been two chess games exactly alike, so again, in the wisdom and omnilaterality of Zarathustra, we have an unlimited variation of possibilities. You can do the same thing with the music of Zarathustra, as I mentioned. Replace the names of the figures as they are moved in any chess game— either from a book or one you play yourself—with the original names, and you have a meaningful, beautiful symphony—a musical description of the interplay of the cosmic and natural forces as we work with them and through them in our daily lives. *(in response to a comment)* Yes, I remember in one of our Seminars held in San Diego, Dr. Garry White, who is the head of the Essene Church there, played the musical accords from my book, *The Art of Asha,* on his violin. I ordained him forty years ago, and he does an excellent job, as well as being a fine musician. I think he not only read all my books, but he read also all those books I didn't write.

So this is the Art of Asha. It is omnilateral. It can be applied to any sphere of life. It is not only beautiful and meaningful, but practical and logical. And in life, only logical things work.

*Why is man described as a combination of a circle, square and triangle—of wisdom, power, and love?*

Well, that is the concept, the basic intuition of Zarathustra. He considered that the symbol of wisdom is a circle without end— wisdom is unlimited, it has no end, it is perpetual. Power is a square. In antiquity whenever they built a tower, or the wall of a city, they used square blocks. Even today the tower survives in modern chess, and it is in exactly the same place as the figure of Power in the original Art of Asha. We still have the expression, when describing someone who is powerful, of being a "tower of strength." Love is a triangle because it represented the trinity. You will excuse my pathological sense of humor, but if you ask a Frenchman why love is a triangle, you will have a different answer: "toujours l'amour. . ." But in ancient Persia the concept was that the triangle is the unity of two opposing forces and the synthesis of the creation of the two opposing forces. What it means is this: There is one opposing force, then another opposing force—then whatever will happen in the world will not be according to one force, or the other opposing force, but whatever will come out from the battle of the two. It is like physics—a parallelogram of forces, and then you have the resultants, and the composing forces, one force goes to one direction like this, the other comes that way, then the movement will not be one way or the other way, but between the two. This is a kind of basic cosmic principle, and in man all is united: power, love, and wisdom, provided man cooperates with the Kingdom of Light. Otherwise, if he is not watching his individual inventory, instead of power he may easily represent weakness—instead of love, hatred—and instead of wisdom, ignorance. This has happened many times in history—a great number of dictators and violent people in all ages who stop collaborating with the forces of Light, and then become a black circle and a black triangle and a black square—hatred, ignorance, and weakness. Yes, weakness. You shall remember one very important thing according to Zarathustra and the *Zend Avesta:* the kingdom of Ahriman is great—we have only to think of the wars, violence, persecution and ignorance on our planet today—but it is limited; it always ends badly. Look at Hitler, for instance, and Napoleon, exiled to Elba. The forces of Light are realities, and they are eternal; but the forces of darkness have only as much reality as we give to them.

*Were the Essenes the last people to practice these concepts?*

Yes, the Essenes were the last who practiced it in its totality. Some traditions survived in a limited way through the teaching of Freemasonry, through the Sephiroth Tree of the Kabalists—but the Essenes were the last ones who practiced the totality, the all-sidedness of the Zarathustrian teaching and way of living.

*What about St. Francis?*

In his beautiful poem, the *Hymn to the Sun,* when St. Francis spoke about Brother Sun and Sister Water and several of the other natural forces of the *Zend Avesta*—which were called Angels'by the Essenes—he simply came into harmony with this wave-length of the ancient Essenes. He intuitively renewed their teachings—like a kind of lightning in a dark medieval sky. We may say that St. Francis was the last Essene.

*Is the wisdom of ancient America directly related to the Essene teachings and to Zarathustra?*

The precolumbian philosophy was a parallel development, not a straight-line development.

*How did the Essenes die out?*

Everything in the universe appears and disappears—nothing

remains as it is. Now the Essenes did exist for centuries around the Dead Sea, and their historical role really was to preserve the ancient Zarathustrian traditions. They wrote down these traditions in the Dead Sea Scrolls, and they had in their library a rich collection of ancient fragments from the *Zend Avesta,* as well as ancient traditions coming through Chaldea, Babylonia, Assyria, Persia, etc. Their role was to keep up the torch through those centuries, to keep them alive and to write them down. The Dead Sea Scrolls were part of the Essenes' library buried at Qumran. *(to Norma)* You may read a few quotations from the Dead Sea Scrolls that concern the sixteen forces. There are some at the end of *The Essene Origins of Christianity.*

*(Norma reads)*

I thank Thee, Heavenly Father,
because Thou hast put me
at a source of running streams,
at a living spring in a land of drought,
watering an eternal garden of wonders,
the Tree of Life, mystery of mysteries,
growing everlasting branches for eternal planting
to sink their roots into the stream of life
from an eternal source.

And Thou, Heavenly Father,
protect their fruits
with the angels of the day
and of the night
and with flames of eternal Light
burning every way.

I am grateful, Heavenly Father,
for Thou hast raised me to an eternal height
and I walk in the wonders of the plain.
Thou gavest me guidance
to reach Thine eternal company
from the depths of the earth.
Thou hast purified my body
to join the army of the angels of the earth
and my spirit to reach
the congregation of the heavenly angels.

The truth is born out of the spring of Light,
falsehood from the well of darkness.
The dominion of all the children of truth
is in the hands of the Angels of Light
so that they walk in the ways of Light.
The spirits of truth and falsehood
struggle within the heart of man,

behaving with wisdom and folly.
And according as a man inherits truth
so will he avoid darkness.

I have reached the inner vision
and through Thy spirit in me
I have heard Thy wondrous secret.
Through Thy mystic insight
Thou hast caused a spring of knowledge
to well up within me,
a fountain of power,
pouring forth living waters,
a flood of love
and of all-embracing wisdom
like the splendor of eternal Light.*

*When were the Scrolls found?*

More or less around 1944. Do you know who found them? They were found by a goat—not by a bear, like me, but by a goat! A Bedouin was pasturing his goats when he noticed that one was missing, and he finally found him in a cave chewing some strange things that turned out to be manuscripts. He grabbed the goat, disciplining him and sending him back to the herd, and picked up the half-chewed thing and sold it to a traveling merchant. The merchant sold it to a little store in Jerusalem, and the store owner sold it to the university, and finally the tremendous importance of the finding was discovered. Here was an ancient manuscript, old as the Old and New Testaments, throwing light on a whole era of mankind. They started to investigate further—a good number of

---

*from The Book of Hymns, the Thanksgiving Psalms, and The Manual of Discipline, all quotations from the Dead Sea Scrolls found in *Teachings of the Essenes, from Enoch to the Dead Sea Scrolls,* by Edmond Bordeaux Szekely.

people descended on that area where previously only goats had pastured—and they did a lot of excavation, discovering more and more things. It was interesting, that in view of the very dry desert climate, these scriptures remained intact—in a temperate climate, or any other climate where there is normal humidity—they would have all been lost. This also shows some wisdom of the Essenes, they perfectly survived in that dryness. Of course, it also had a disadvantage: these manuscripts were all wrapped in linen, and the linen was put in earthenware pots, and these pots were buried in caves. And when they unwrapped the linen, they found the manuscripts extremely brittle, and they all fell apart. Then they had to put them on a long table and sort everything like a huge puzzle—a very time-consuming and difficult job. Then pictures were taken from above and the photos sent out to universities and theological seminaries, and so on. And it all started with that most illustrious goat who wandered away from his Bedouin shepherd and found the first Dead Sea Scroll.

*Were the linens soaked in oil?*

No. The oil would have decomposed and caused putrefaction, destroying all the scrolls.

*Did you do any work in translating these?*

Yes, I did translate a lot of the Dead Sea Scrolls. I used several of them as mottos for each chapter of my book, *Teachings of the Essenes, from Enoch to the Dead Sea Scrolls.* In fact, it will be a wonderful way to close today's session, if Norma will read one of them, that begins, "The Law was planted in the garden of the Brotherhood. . ." The whole philosophy of the Essenes, based on the ancient traditions of Zarathustra and the *Zend Avesta,* is found in this beautiful excerpt from the Dead Sea Scrolls.

> The Law was planted
> in the garden of the Brotherhood
> to enlighten the heart of man
> and to make straight before him
> all the ways of true righteousness,
> a humble spirit, an even temper,
> a freely compassionate nature,
> and eternal goodness
> and understanding and insight,

and mighty wisdom
which believes in all God's works
and a confident trust in His many blessings,
and a spirit of knowledge
in all things of the Great Order,
loyal feelings toward the children of truth,
a radiant purity which loathes everything impure,
a discretion regarding all the hidden things of truth
and secrets of inner knowledge.

Praise be to Thee, O Lord,
For our mother, the earth,
Who sustains and nourishes us,
Bringing forth diverse fruits,
Flowers of many colors,
And the grass.

—St. Francis

## The Fourth Day: July 29, 1979

During the first days of the Seminar we discussed the foundation of the Essene Teachings, the *Zend Avesta* of Zarathustra. Now I want to go into detail regarding the science of Biogenics—exactly that science which tells us how to apply these beautiful ancient teachings in our daily lives in the twentieth century. This is extremely important, because without knowing how to apply them in our daily lives, all these beautiful teachings of Zarathustra and the Essenes will be like a gold mine on the moon—very great value, but not much use to us.

I want you to realize that not only we, but all living beings, from the most microscopic to the most gigantic, are living in the biosphere of the earth. The biosphere is that area which goes deep down into the earth, many hundreds of feet, and thousands and thousands of feet up into the atmosphere, and of course includes the surface of the earth where all the species on our planet exist. This is the biospheric zone of our planet, where life has its thermic and other preconditions, where organisms can adapt themselves adequately to the environment. But in this immense biosphere which surrounds us, there is a much smaller zone—those few inches of top soil, and the few inches above it, where all the life forces and elements necessary for life meet—this is an intensive life-generating zone, where life is continuously being created. In spite of our tremendous accomplishments in technology, we would perish if these few inches of top soil would suddenly disappear. We have lessons in history, such as the great precolumbian civilization of the Mayas on the Yucatan peninsula. They didn't read Zarathustra about natural gardening, and planted corn, and corn again, and corn again, until finally they exhausted all the fertility of the soil. They just took and took from the soil and gave back nothing, finally arriving to a stage where, without any natural catastrophe, without any wars, they just disappeared, because the fertility of their soil disappeared. *Historia est vitae magister,* says the Latin proverb—history is the teacher of life. In certain diabolical ways our present technological civilization is very efficient, and we are again on our way, through the inundation of our planet of chemical fertilizers, insecticides, defoliants, etc., to destroy our top soil.

This zone around the fertile top soil is the *biogenic sphere* on our planet, created and developed by nature through millions and millions of years. It is our greatest treasure. There are all kinds of organisms living all over in the larger biosphere, but *life generates* in this biogenic sphere. There is something—a biogenic vitality in the top soil and in the immediate surrounding atmosphere—this is biogenic power—life-generating power. You will be very surprised if I tell you that the main historical event in the history of our planet was not a great battle, or the birth of a great genius, or the discovery of the airplane or rockets, but something that happened fifty million years ago which changed completely life on our planet: the appearance of grass. This simple, humble grass—came to prepare our planet for the advent of man, who came many, many millions of years later. Without this simplest of plants, grass, there would be no animal life on our planet, and of course no humans could have appeared.

This life-generating power is biogenic energy. Everything which is generated, be it a plant, a microorganism, an animal, or a human being, first has a *biogenic* period of life, immediately following birth. When a plant is born from a seed, there is an outburst of biogenic energy—it is surrounded by it—and the plant starts to grow with tremendous speed and intensity. Look, for instance, at this wheat grass growing. You plant a grain of wheat and in five days it is several inches tall—you can practically see the great speed of its growth—this is the biogenic period of the plant. Later on, the growth slows down. The plant is still active, still growing, but has only enough energy to insure its own health and growth—none in excess. This is the *bioactive* phase of the plant. The third stage is entered when the plant becomes more static, growth is insignificant—some leaves become yellow, some deficiencies develop, and the biological functions slow down and become static. This is the *biostatic* phase. And finally the plant is decaying. This is the fourth stage of life, the *biocidic* stage.

A very interesting technique was developed in Russia—they are sometimes doing some good things here and there—and that is Kirlian photography, which is able to photograph the forcefield around a plant. During the first biogenic period of this exuberant fast growth the photographs show a very intensive and large forcefield. In the bioactive phase when the growth slows down, the photographs show a less wide forcefield, and less dense. Still later, in the biostatic phase, the forcefield is very weak and pale, and when the plant starts to decay, disappears completely.

Now this is an extremely important biological law. I always wondered why it was not discovered. This is the greatest, the most important law of life, these four ages—the *biogenic,* the *bioactive,* the *biostatic,* the *biocidic.* It is so evident, it is a universal law. It happens also with the human organism. You can see a little baby growing very, very fast—it is the biogenic exuberance. Later you see the adolescent—still growing, but the speed of growth is not so fast—it is the healthy bioactive stage. Then we reach the age of maturity. The vitality is not the same as a baby or an adolescent—we are functioning all right, but the growth, the exuberance of the previous ages is not there anymore. And finally, the biocidic age appears when we get very sick and die. It sounds a little like Buddha! I think it was a great physiologist who said that the processes of death are starting with birth—we are creating every day billions of new cells, and we are destroying billions of cells in the form of catabolic waste products.

Every animal organism, every human organism, every tree, every plant goes through these four stages. Shakespeare gave a fine description of the ages of man, as did the Brahmins. But scientifically it was completely neglected until the appearance of my book, *La Vie Biogenique*—Biogenic Living.

The science of Biogenics also classified the foods we eat—now here comes the practical importance of this scientific concept—based on this universal biological law, we classified foods in four categories: biogenic foods, bioactive foods, biostatic foods, and biocidic foods. *Biogenic foods* are life-generating—foods in which biogenic vitality erupts in exuberance, like sprouts. For example, wheat germinates—then suddenly starts to grow with tremendous speed. And this is not only an appearance, but profound biological processes are going on in the biogenic phase of a plant which can be measured. For instance, the Vitamin C content of wheat increases 600% when it is sprouted. The Vitamin C in soy beans increases 500% after sprouting only a day and a half, and Vitamin B2 increases 1300% in oats after sprouting, and so forth. Reading reports and research from different universities, you can realize what is going on in the biogenic phase of a living being, easiest to observe as a seed germinates, as it sprouts—we can see and feel and measure all the characteristics of biogenic exuberance. And this is not just theory, but we can support it with facts, all those things which are happening to the vitamins during the biogenic period.

Here are laboratory reports, as described in my book, *The Chemistry of Youth*—I want Norma to read it to you.

*(Norma reads)*

—Dr. Ralph Bogart, *Kansas Agricultural Experimental Station,* sprouted oats and found in a quantity of 40 grams, 15 milligrams of Vitamin C, more than in the corresponding amount of fresh blueberries, blackberries, or honeydew melon.

—Dr. Berry Mack, *University of Pennsylvania,* found that his sprouted soya beans by the end of 72 hours had a 553% increase in Vitamin C.

—Dr. C. Bailey, *University of Minnesota,* found only negligible amounts of Vitamin C in wheat, but after a few days of sprouting, he found a 600% increase.

—Dr. Andrea at *McGill University,* found 30 milligrams of Vitamin C per 110 grams of sprouted dry peas, favorably comparable to orange juice.

—Dr. Beeskow, *Michigan Agricultural Experimental Station,* found the maximum of Vitamin C in sprouts after 50 hours of sprouting.

—Dr. Paul Burkholder, *Yale University,* found the Vitamin B2 content of sprouted oats increased by 1300%, and when the little green leaves appeared on the sprouts, the amount increased to over 2000%. He also found the following approximate increases in:

> Biotin 50%
> Inositol 100%
> Pantothenic Acid 200%
> Pyridoxin (B6) 500%
> Nicotinic Acid 500%
> Folic Acid 600%

—Dr. Francis Pottinger Jr., from California, found sprouted legumes and beans to contain first quality, complete proteins.

—Dr. Clive McKay, *Cornell University,* wrote a series of articles recommending a "kitchen garden" of sprouts in every home to produce fresh sprouts through the year.

*(Professor continues)*

As you can see, these basic facts support one hundred per cent our biogenic philosophy, that during the biogenic phase of any plant, intensive biogenic processes are going on, increasing tremendously a multiplicity of nutritional values in the plant. Therefore, it is evident that they represent the most potent and active foods for us, especially for therapeutic reasons. I found, during my more than fifty years of medical practice, that when we have a very serious case, when we have little time to act and must get the maximum therapeutic results, then we use one hundred per cent biogenic substances.

*Bioactive foods* are those which are excellent nourishment, fine foods—but they are already fully developed—like fresh raw fruits and vegetables—they are excellent for nutritional purposes, but these are bioactive foods, and in their body we have no more this spectacular outburst of multiplication of vitamins as we have in the biogenic sprouts. Nevertheless, they are excellent nourishment; they have plenty of vitamins, minerals, enzymes, etc., although in lesser quantity.

Then we have the category of *biostatic foods*. These are not fresh foods, but they still furnish minerals and calories, which you need to function—these are all the different foods which are stored, which are non-fresh. We found that a very well-balanced diet for daily living is 25% biogenic foods, 50% bioactive foods, and we can afford the luxury of 25% biostatic foods.

But we have a fourth category: *biocidic foods*—foods which contain chemicals. You have no idea the thousands of chemicals that appear every year—the preservatives, the coloring substances, the anti-oxidants, the humectants, the emulsifiers, on and on without end. With this category of biocidic foods we shall not make compromise. We can afford to eat 25% biostatic foods, but never biocidic foods.

*Professor, what does it do to a food to quick-fry it in a wok?*

To quick-fry in a wok is a definite improvement over the deep-frying process of American "civilized" foods, but of course they will not be anymore bioactive foods. They will be somewhere between bioactive and biostatic. But the other kind is definitely biostatic.

I want you to have an idea about biocidic foods. I have a page in my book *The Chemistry of Youth* where I give a condensation, just a drop in the ocean, of the chemicals being used in the food industry today. Please listen to this flood of chemicals:

*(Norma reads)*

After the second war, an accelerated flood of synthetic and toxic additives inundated our markets and supermarket chains, which soon became all-sidedly omnipresent in *all* products which the innocent and ignorant housewives carried home to deteriorate the health of their families. Instead of purity, freshness, and wholesomeness, the new post-war criteria for desirable food became taste, texture, and shelf life. In geometrical progression ever since, the greediness of the lethal food industry and the all-pervading sophisticated and deceiving promotion of television and radio have become an omnipresent menace of corruption to the public mind and body.

It is mind-staggering to try to conceive that, according to the statistics of our bicentennial year, 550 different synthetic chemicals, a total of a billion pounds a year in over 32,000 products, are conspiring against the innocent, uninformed, and misinformed consumer, making it very improbable that our nation will ever survive to celebrate a tricentennial anniversary.

We will mention only a few infinitesimally small categories of these synthetic chemicals: preservatives, emulsifiers, moisturizers, dyes, sprays, bleaches, artificial flavors, gases, antioxidants, hydrogenators, deodorizers, buffering agents, alkalizers, disinfectants, acidifiers, extenders, fungicides, insecticides, drying agents, defoliants, thickeners, neutralizers, conditioners, maturers, antifoaming and anticaking agents, artificial sweeteners, fortifiers, hydrolizers, etc. etc. etc.

Whenever food manufacturers remove a natural substance from a food, they always replace it with a synthetic adulterant. Of course, this manipulation disturbs the whole natural biochemical balance in the food. Each time a synthetic is consumed, the biogenic and biological processes of the organism receive a shock to which they are desperately trying to adjust themselves, very often with little success, Each time, a precondition of disease is added to the already disturbed living processes, developed and perpetuated over millions of years. Their continuous, more and more aggressive disturbance proliferates a long line of chronic and degenerative ailments. Even in ordinary simple foods, a great number of synthetic, and often toxic additives are ingested.

In ice cream: coal tar dye, diglycerides, monoglycerides, antibiotics, artificial flavors, carboxymethyl cellulose, artificial colors, etc.

In apple pie: nicotine, lindane, chlordane, parathion, demeton, lead arsenate, methoxyclore, butylated hydroxyanisole, benzene, hexachloride, sodium-phenylphenate, malathion, etc.

In butter: diacetyl, hydrogen peroxide, coal tar dyes, etc.

In oleomargarine: monoisopropyl citrate, diglycerides, isopropyl citrate, monoglycerides, etc.

In pickled vegetables: sodium nitrate, alum, aluminum sulphate.

In fruit juices: saccharine, parathion, dimethyl polysiloxane, benzoic acid, etc.

In breads: coal tar dye, diglycerides, ditertiary buthyl paracresol, ammonium chloride, polyoxyethylene, monoglycerides.

In meats: stilbestrol, dieldrin, aureomycin, methoxyclor, heptachlor, toxaphene, chlordane, benzene hexachloride, etc.

In potatoes: all kinds of pesticides: ethylene dibromide, dieldrin, heptachlor, chlordane, etc.

We could continue on page after page the endless thousands of synthetic and toxic additives, but I think these few samples are enough. *Sapienti sat!*

*(Professor continues)*

As I mentioned, this is only one drop in the ocean. You have no idea the thousands of new chemicals that are thrown into the commercial products of the food industry every year. Of course, when finally the government discovers that many of these are lethal to health and carcinogenous, then after a great legal battle they prohibit the use of one; but meanwhile, thousands of others appear and the burden of proof is on the government—they may catch one-tenth of one per cent of these harmful substances, but it is impossible to stop all of them.

*What about the influence of chemical additives in the foods here in Costa Rica?*

Up until fifty years ago, everything was natural here, everything. Then the wonderful American food industry discovered, oh there is a fine market in Central America. So they established all kinds of fast-food chains, like the Kentucky Fried Colonel, Dairy Queen, McDonalds, etc., and they started to brainwash people in Latin America that they must keep up with progress in the world and not stagnate with the old foods that were eaten in the past, but just follow the American way of life and have all these wonderful new things: soft drinks, packaged and canned foods, chemically-processed meats and cheeses, etc. etc. With diabolical speed and efficiency they began the wholesale poisoning of the people of Latin America. It worked out well for them, because when the government once in a great while is able to prohibit the use of something which is evidently carcinogenous, these companies may have a tremendous storage of millions of pounds of that substance, and they cannot afford to destroy them, so they sell them to small countries who do not realize what is happening. So this is our civilizing influence, you see, in Latin America. Unfortunately, fewer people in Latin America are reading the writings of Thomas Jefferson and Lincoln and Walt Whitman and Edgar Allan Poe and Emerson, and thousands of times more are getting the non-literary benefit of all these chemicals.

Now, going back to the mainstream of things: these are what I call biocidic foods. Please remember that if you eat 25% biogenic foods, like sprouts and fast-growing baby greens, and 50% bioactive foods, such as fresh fruits and fresh vegetables, you can afford to have 25% biostatic foods, which are not fresh but still good, as they don't contain chemicals or other harmful substances, and you will have a fairly good nutrition. But beware of biocidic foods. Yet it is extremely difficult to avoid them. In fact, things are getting worse and worse. At the end, the only way to avoid them completely is to grow a few basic foods in your back yard, even in your room—we will discuss this next—how you can grow a lot of things even if you live in an apartment in the city, where we suppose there is no ground, no soil. Even so, there are hundreds of sprouts you can grow, and small greens, such as onion greens, garlic greens, beet greens, lentil greens, etc. We will discuss it later in detail.

For the moment, I want you to realize and remember well these four categories of foods. It is not enough to classify foods according to their mineral and vitamin content—*we must classify foods according to what it does to us after we have eaten it*—not just whether it is a protein food or a starchy food, or how many minerals are in that can or package, but *what is happening to it after we ate it.* This is extremely important—I had to reform the whole system of nutrition and diet in the biogenic way by classifying foods according to these four categories of biogenic, bioactive, biostatic and biocidic foods. For the most important thing is the *dynamic action:* what is happening in our organism after we are eating the foods. Now if we go only by the mineral or vitamin value of a food, well, that is very nice, but at the same time that food can have biocidic elements—it can have a dozen chemicals—no matter how high the vitamin and mineral content, it will not be a good food.

There are foods which create a biogenic state in our organism, with great exuberance of life powers. Then there are foods which are maintaining our life very well. Then there are foods which are gradually and progressively creating old age. And finally there are foods which are creating death—the biocidic foods. Now when we are born, we inherit a certain capital of vitality, like inheriting a bank account. Everything depends on how we spend that bank account. You can spend it very fast and become old at the age of thirty and die at the age of thirty-five. But that vitality, that capital of vitality we inherit, if we handle it with prudence, if we don't spend it unnecessarily by wrong eating and living habits, if you will not drink alcoholic drinks, if you will not smoke, if you don't inhale polluted air and drink polluted water and eat polluted foods and introduce chemicals into your system, then your youth will last much longer, your adolescence will last much longer, your age of maturity will last much longer, your old age will last much longer, and you will increase tremendously your longevity. But here comes into play these four categories of foods: biogenic, bioactive, biostatic, and biocidic foods. This is a safe guide: you don't need any one-sided or complicated nutritional system with theories which have no scientific foundation—here are the *four basic realities of nutrition*—these four categories of foods.

*What about the seeds we plant—do they contain chemicals?*

Well, there are seeds which contain chemicals, yes, because the seed industry's main purpose is to make money, not so much to take care of your health. But there are some reliable seed companies—you have to do some detective work and you will find them. Sanity is returning in homeopathic doses.

*Norma: Professor, I would just like to say that often when we choose those fruits and vegetables which are native to the land where we live, they have a much better chance of being grown without chemicals. For instance, the potato is not native to Costa Rica, and all kinds of chemical fertilizers and insecticides are needed to raise it to maturity. But the humble* tequisque, *very similar to the* taro *of China, and even more nutritious than the potato, grows wild here without any help at all.*

Yes, that is very interesting about the *tequisque.* When I studied precolumbian archeology there is a very beautiful book by a Franciscan monk who came with Cortez, Fray Bernardino de Sahagun—the book is called *Historia General de las Cosas de Nueva España*—in which he describes all these ancient precolumbian fruits, vegetables and roots—(although at that time they were not ancient; everyone ate them!)—and among them were *tequisque, ñampi, pejivalle,* and similar things. And to my great surprise, I found when I went to Mexico to do archeological fieldwork that these foods no longer exist! No one knew what was *tequisque* or *ñampi* or *pejivalle.* Then, when I came to Costa Rica for the same archeological purpose, I discovered that all these foods which are extinct in Mexico are growing wild here, flourishing all over, sold in the market, etc. So it seems that tiny Costa Rica has done better than Mexico, which is much closer to the United States—there the ravages of the food industry are the worst. They have successfully exterminated not only *tequisque,* but also most of the other wonderful precolumbian fruits, vegetables, and roots. It reminds me of what Porfirio Díaz said—he was dictator of Mexico for forty years and was finally overthrown by the Mexican revolution. He got on a boat in Vera Cruz and went to France, and when he left, he said, "Pobre México! Tan cerca de los Estados Unidos, y tan lejos de Dios!" (Poor Mexico—so close to the United States, and so far from God!) And I think that statement is still valid today.

I don't know—I very often don't mention that I am coming from the United States when I am asked—I simply don't feel to be very proud of what we are doing in Latin America concerning public health, and fertilizers, and defoliants, and insecticides, and all these things.

*Professor, what about hybrids?*

A great many new species are wonderful, excellent. But the one disadvantage of hybrids is that you cannot use their seeds to plant new vegetables—you must use seeds of non-hybrids. Remember I spoke about Luther Burbank, whom I consider the Zarathustra of the twentieth century—he created about two dozen varieties of apricots containing all kinds of different enzymes and vitamins—delicious, wonderful foods—but don't depend on the seeds, because they will not work. They are not self-producing. So if you live in a place near a city, where you can always get seeds, fine, you can have all kinds of hybrids. But if you live far away where it is difficult to get all these seeds, then you'd better use non-hybrids, because then you can save the seeds from each crop and perpetuate them.

We discussed the sad state of nutrition all over the world. Almost fifty years ago, when the situation was not so bad as today with the tremendous amount of unnecessary chemicals in foods, I reached the conclusion that the only way you can be absolutely sure your foods are fresh and have nothing wrong with them, is to grow your most essential foods yourself, on a very small scale. I remember during my summer seminars in England in the early thirties, one of the participants came to see me after one of my lectures—he was E.F. Schumacher. He told me, you know, I read yesterday your book *Cottage Economy*—you are turning upside down the economy—it is a completely new system of economics. He was working at that time in international organizations, and later with the United Nations, trying to inject these ideas, my philosophy of the greatness in the smallness. I persuaded him to try his very best, but he was not very successful—it was not easy to go against the growing food industry, government, politics, etc. Nevertheless, twenty-five years later, when he retired from his public activities, he wrote his book, *Small is Beautiful*, advocating a kind of intermediate technology, between the small and the big. Well, in that book, *Cottage Economy*, and later in *Father, Give Us Another Chance* and *The Greatness in the Smallness*, I advocated a system which virtually anyone can use to grow all the essential foods they need in a very small area. Few people realize how little space is needed to grow foods—the richest foods in vitamins, minerals, enzymes—an extremely small area. Now I am traveling most of the time—south and north and east and west—and I never stop anywhere long enough to harvest a crop or to plant one. Nevertheless, I follow myself the ideas I recommended to the British readers in my *Cottage Economy*. The first thing we do, wherever we are, is to sprout seeds. Sprouts can be harvested in four days, you don't need any large areas, you need only two square feet and a few jars. Then, it is necessary to have a supply of biogenic, fresh, chlorophyll-rich and vitamin-rich greens. So we always make a few wooden boxes—you can see some of them here.

We plant in the boxes green onions, which can be harvested in three weeks; garlic greens, which can be harvested in less than twenty days, and so on. You can plant a mature beet, and it will give you tiny beet greens almost right away. You can sprout lentils and spread them over the soil in the box and in one week you have a good supply of lentil greens. There are many kinds of greens you can grow very fast in these little boxes and have a kind of instant garden. If you look at that corner in the shed beside the vegetable garden, you will find a lot of boxes thrown out which were already harvested, finished a long time ago, and new greens have started to grow again! It shows you how it is possible in a very small area to grow a lot of fresh foods. And if you have a small garden, only a few square yards, you can really grow a tremendous amount, by planting quick-growing foods which have the maximum amount of biogenic vitality. So there is the greatness in the smallness. I say it again, because I want you to remember—it is possible, wherever you are, in the minimum of space, to grow your most essential foods and have your real vitamin and mineral supplements growing in your miniature garden.

Now in our family, it is Norma who is taking care of growing our baby greens and sprouting our seeds, so maybe I will let her explain to you the practical way of doing it.

*Note from Norma: The explanation covered the basics of sprouting seeds and growing baby greens, most of which is familiar to the reader. For all details, please see Professor's books,* The Chemistry of Youth (Search for the Ageless, Volume III) *and* The Essene Way-Biogenic Living.

### (Professor continues)

Now it is much more important to do this in a practical way than it was forty-five or fifty years ago. At that time, the food industry didn't contaminate 80% of the world's food supply, as it is doing today. Fast-growing plants don't need water—they need moisture. The essential thing is moisture. Now of course, if you are living in the dry air in the city somewhere, you have to give your plants a few drops of water. But here I want to inject a very important principle in gardening which was perpetuated by the ancient Essenes at the Dead Sea.

We had a Roman natural scientist, Plinius, who had a true Roman empirical mind. He didn't care what kind of philosophical or theological ideas some strange people in the middle east might

". . . this is the real greatness in the smallness—a microcosmos, a miniature thing which embodies the whole cosmos, all the sixteen forces of the Creation, all in this little biogenic battery." *(p. 166)*

have, but he was intrigued by rumors which came to Rome that around the Dead Sea, in the middle of the driest desert in the world, far away, nearly inaccessible from an inhabited area, with no water supply, there was a community which was growing fruits and vegetables and luxuriant trees. Plinius, with his empirical, investigative mind, was fascinated, and he undertook a trip to Palestine to have a discussion with the Roman Procurator—the governor—about the possibility of visiting the community. The Procurator offered him a Centurion and one regiment of Roman soldiers to take him to that area. Plinius thought about it, but after meditating for about a week, refused the offer. He was thinking, well, very fine, there will be a safe escort. But these soldiers are always marching in full armor—what will happen to them in the hottest desert in that armor—he was sure they would be liquidated on the second or third day, and then he would be left alone in the desert without any help. So he thanked the Procurator but said he would try to find some other way to get there. He made the acquaintance of a lot of natives, gradually getting more and more information about that strange community at Qumran, and finally he successfully persuaded one small group who themselves wanted to visit that community, to take him with them. It turned out to be a fine idea, because they explained to him that there would have been no chance of his arriving with that cohort of Roman soldiers. Even if they would have been able to survive the desert and not be baked in their hot armor, even so, the people of that community would not have looked very happily at Roman soldiers arriving at that place—because the first reason why they secluded themselves in that remote area was to escape the Romans, and escape taxes, and escape suppression. They surely would not be eager to tell him anything about their methods of growing plants if he were to arrive with Roman soldiers! So he went with that small group, and that was how Plinius made contact with the Essenes. He made observations, and little by little gained their confidence, and explained to them that although he was a Roman, he had nothing to do with the Roman government, he had no harmful intentions, he was only interested in their method of gardening. How was it possible that they were growing foods and trees and plants and everything without water? And finally, after observation and meditation, he discovered the answer. That area—where the Dead Sea Scrolls were excavated—is an area surrounding a dead sea. From the Dead Sea the water did evaporate for thousands and thousands of years, and as more and more water evaporated, the remaining water got more and more saturated with salt—for the salt did not evaporate—and the water was impossible to use for gardening. Now during the day, the heat is inhuman. When we did our excavations for the Dead Sea Scrolls, we were sleeping during the day—we did not make any attempt to go out in the sun and work. We did our work at night. It doesn't mean the night was cool, but it was not too hot. So we slept during the day and were working at night. (In fact, I got into that funny habit, and since then I am writing my books at night and sleeping in the morning—habits grow on you.) Now the heat of the Dead Sea was just as intense for Plinius, and he observed that under the hot sun a large amount of vapors were continually rising up into the atmosphere, because the tremendous heat of the sun made evaporation very intensive and continuous, and once the vapor rose, saturating completely the atmosphere, tuen the sun went down. Now if you have traveled in the Sahara or that region, you probably noticed that as soon as the sun goes down in the desert, for several hours it becomes extremely cold. So what happened—that suddenly cold air precipitated the vapors which were up in the atmosphere, and the whole thing came down as heavy dew. You have probably noticed in your garden in the early morning, little drops of dew like pearls on the leaves, without any irrigation. But in the desert the dew formation was much heavier, more intensive. And it was that dew that the Essenes used to grow plants and trees and everything. According to the description of Plinius, they created beautiful luxuriant gardens and orchards and produced all their foods, without irrigation, using only the dew.

Plinius described another very ancient system, long forgotten, but these days resuscitated by organic gardeners, by the same empirical method: the use of mulch. The Essenes used dry leaves and all kinds of other substances to cover the ground, and with this mulch broke the process of capillarity. Through dryness in the air, through the action of wind and sun, a lot of moisture is lost from the ground—through capillary action it rises into the air and is lost from the top soil. If you realize this method of the Essenes you can have a garden with one-tenth of the water you may usually need. You collect the leaves that fall from trees, the grass you cut, even pieces of paper, and cover the ground with them. These will break the capillarity, and the moisture underneath will remain intact and not evaporate. This was one of the methods used by the Essenes: the conservation of moisture. They were unable to get water, but they were able to conserve the moisture they did get from the heavy dew.

A lot of people today think that organic gardening is something new and revolutionary—well, not at all. Here we have Plinius' beautiful description of the Essenes' methods, and much earlier we read Zarathustra in the *Zend Avesta* describing it as a natural, normal method of gardening. In fact, I may say this: it is the non-organic gardening that is something new, not organic gardening. Since the chemical industry of fertilizers, insecticides, and all those related chemical products came into existence, farmers and gardeners have been brainwashed into thinking that this is the only way of gardening. Then the ideas of organic gardening had a renaissance—first by Sir Albert Howard, who practiced it in India, and then propagated these ideas. Of course, he didn't read the *Zend Avesta* of Zarathustra, but by common sense he arrived to this conclusion, that this is the right way of gardening. Then suddenly the idea caught on, organic gardening appeared, and people accepted *de facto* that organic gardening is a kind of new fad—it may work sometimes and may not work—but there was nothing new in it—it is thousands of years old. It is the gardening with chemicals that is new.

*Did the Essenes let the dew fall on the mulch, or did they move the mulch and let the dew feed the ground and then put the mulch back?*

With very tender plants, small plants, they removed it so the plants could grow well. But otherwise, they let the mulch remain, because the lower part of the mulch gradually did decay and did enrich the soil. This is the principle of the Essenes—that we must put back into the soil what we are taking out. So the lower part was always gradually decaying. But they always added some dry substance on the top to break the capillarity, so that there would be no loss of moisture by the capillary action going upward and dissipating into the air.

One argument of the chemical agriculture industry for gardeners is that small farms are impossible—they are not profitable because they have no money to establish methods of irrigation. Of course they are also selling irrigation equipment—sprinklers and tubes and all the accessories. But none of that expensive irrigation equipment is necessary if you practice the principle of conservation of moisture. *Plants don't need water—they need moisture.* This was the basic principle of the Essene method of gardening, and the basic idea expounded by Zarathustra in the *Zend Avesta.* It is a principle practically forgotten in the twentieth century, and one that is not very popular with the companies selling irrigation equipment. But I want you to remember always this ancient, tested, and proven idea of utilizing the dew of the night from the atmosphere, and then trapping the dew downstairs in the topsoil by breaking the capillary action through the use of dry mulch. This is the only sound method of preservation and conservation of moisture.

Now if you live in the city and are growing greens in little boxes, naturally you will have great difficulty in capturing the dew, especially if you have an apartment on the tenth floor. In that case, you cannot follow the Essene methods 100%—you have to give a few drops of water from time to time. But if you live in the country and have a small garden, just a few square yards, then you can beautifully practice this method. But if you are too close to a large city where the atmosphere is polluted, then even the rain water which falls will be polluted because it absorbs the chemicals from the air. So you can practice this method only when you are away from the large cities.

Another point I want to make—here again the greatness in the smallness—do not underestimate the quantity of greens you will harvest from those little boxes. When I wrote *Cottage Economy* I was weighing the amount of greens produced through one day in a simple little box like this—if you had a dozen you could export the excess! So really you don't have to go every time to the supermarket and buy those non-fresh, stale, sinister-looking vegetables waiting for you under florescent lights. With a few simple wooden boxes or buckets you can grow all the fresh things you want right where you are, and plenty of them. And don't forget the sprouts—sprouts can be harvested every four days—that is a continuous intensive production, and you don't even need soil at all!

Sprouting, especially sprouting of wheat, is beautifully described in a chapter, or Vendidad, of the *Zend Avesta* of Zarathustra.

*Professor, what is a Vendidad? I noticed that name, and also Yasna, when I looked through your translation of the Zend Avesta.*

Vendidad and Yasna are simply names of divisions of the *Zend Avesta*—what you would call chapters today. And one Vendidad deals with sprouting. Remember I mentioned to you that Zarathustra used his garden as a school. He taught the laws of nature in his garden to his disciples. He taught them how to sprout, how to grow little greens, how to grow fruit trees, and in the process demonstrated different laws of nature. So you can have a little Zarathustrian garden wherever you live, even in an apartment, because of the principle of the greatness in the smallness. But even though it is small, you will be surprised at the amount of foods you can grow and how often you can harvest.

An old friend of mine, a great Greek philosopher, Protogoras, said, "man is the measure of all things." This is a tremendous statement—think about it. When we want to produce foods, we have to check first what are the needs, the requirements of the body—not what the radio and television and the literature of the big chemical companies are trying to tell you—that your garden must be profitable, that you have to grow so much and so much, the fastest way, the biggest quantity, with chemicals, and so on. Yes, but we do not have requirements in our bodies for chemicals! We have requirements for Vitamin A, Vitamin B, Vitamin D, and so on, but we have no requirements at all for these synthetic chemicals. Therefore, man is the measure of all things, says Protogoras. So we have to see, what are our requirements? Now our requirements are certain substances, like chlorophyll, for instance, which we can grow in four days—the grass in these little boxes and batteries is only four days old. Then there are these garlic greens. We have a requirement for the essential oil in garlic, *oleum allii,* which is a natural penicillin, but without side-effects and after-effects. When we eat garlic in homeopathic doses, little by little, a handful every day mixed with our foods, it has a slow, penicillin-like effect, especially against certain toxic conditions of the colon, destroying harmful bacteria in the system, and it works continuously. Well, chlorophyll is a natural requirement. When we don't eat greens, we have a deficient diet. And *oleum allii* is a very fine requirement for the human body. And you heard before when Norma read to you the reports from the different universities concerning all the biochemical changes, all the exuberant multiplication of vitamins, hormones and enzymes in new, biogenic little plants. Now we have requirements for these things. But we have no requirements for those hundreds of chemicals in that other list that was read to you. Therefore, please always remember my old friend Protogoras—man is the measure of all things. We shall use as our point of departure the question: what are our requirements? Then we shall plant exactly those things for which we have a requirement, and in the amounts for which we have a requirement. Just because something is healthy and good, it doesn't mean that much more of it will be even better. Once we find out what is the optimal amount for us of these sprouts and fast-growing greens, *then we will realize how little we need of these things.* This is an extremely important thing: here again, the greatness in the smallness. Some authors expound through a whole book about the revolutionary idea that you can live very well on only one acre of land, thinking they made a great discovery. You don't need even an acre if you live in the country—a tenth of an acre is more than what you need. And if you live in the city and grow little things, you need only a few square yards, total.

*What about drinking wheat grass juice?*

There is a movement which advocates growing wheat grass, then cutting and pressing the grass and drinking the wheat grass juice. Well, I want you to realize one thing: that in order to drink a cup of wheat grass juice, as they advocate, you would have to use about twenty times as much wheat grass as is here. Now do you have any idea how much chlorophyll and other substances that concentration would represent? Here it is again—they think that if something is good, then a lot of it will be better. But it just doesn't work. Most of the time, people vomit it out very shortly and then have stomach troubles and all kinds of things when they drink such an amount of concentrated juice from the grass. Fruit juices are less concentrated, but some people go ahead and drink a quart of carrot juice a day—I know a number of people who do that—and then they are surprised when they develop the symptoms of jaundice. You see, there is no such thing as the greatness in the greatness—only the greatness in the smallness.

The same people recommend to drink the water that was used to sprout the wheat—they save that water and drink it. But that water contains phytates, which is a kind of natural insecticide covering the whole wheat grain. In this way nature protects the grains in the soil until they can send out sprouts and roots. But it is not meant to be ingested, and when you soak the wheat grain before sprouting, the water will naturally be full of phytates. But of course, the same people—who are very good people and mean well—they do good things also. They tell everyone not to drink alcoholic drinks, not to smoke, not to eat meat, not to eat processed foods, white sugar, white flour, etc.—and some people get results simply by cutting off all these things. But not from drinking large amounts of concentrated wheat grass juice and not from drinking the water that wheat and seeds were soaked in. Remember when I quoted George Bernard Shaw—he said there is nothing worse in the world than ignorance in action. They may have the best will to help their fellow man, but sometimes they do just the opposite.

The Essenes used grass, because grass is growing very fast—but they used it in homeopathic doses. One food they ate which Plinius found very delicious was a mixture of dates and a small amount of fresh grass. That was his favorite food—he liked it very much. You can use it too—it will be a fine addition of vitamins and chlorophyll to your vegetable salad, whole grain dish, or cooked vegetables—but not more than one spoonful, as it is very potent.

*How can you get the symptoms of jaundice from drinking carrot juice?*

The liver is a tremendous organ. It is working incessantly and performing a lot of basic biological functions. But suppose you have a donkey who can carry a certain weight on his back—he may carry that certain load very well, but if you overload him, he will collapse. Now it is the same thing with the liver. If you overload it with certain substances, the poor liver will start to work with great difficulty, and then those substances will get into the bloodstream which distributes it to the tissues, and then you will develop beautiful yellow skin and yellow eyes, and so forth. And it takes some time to get rid of that condition. Here again, remember, there is no greatness in the greatness—only in the smallness. It is very difficult to take an excess of Vitamin C or Vitamin B unless you go really overboard; but Vitamin A and Vitamin D can become toxic if you have too much of them. And, you see, carotene transforms into Vitamin A in the system, and there you can get an excess. And think of one thing: how many carrots do you have to use to make a quart of carrot juice? Does anyone know? All right, six to eight carrots. Now would you sit down and eat six to eight carrots at a meal? You would not. Well, it is evident, just a little common sense is needed.

*Can the use of electric machines affect the quality of the food?*

Well, there is one thing here: a high-speed electric machine is stirring in a lot of oxygen which creates oxidation. Oxidation is harmful—it reduces tremendously the biochemical value of the juice. Therefore, I always said what is wrong with eating vegetables and fruits as they are? Why shall we put them in a machine and oxidize them and drink that juice? I don't see any reason why—because in the process you are losing valuable fiber. In the juice-making process the fiber is thrown out. Of course, it may be very useful in your compost heap—but I think your organism, your colon, and your system needs that fiber more than the compost heap. So I really don't advocate the so-called juice diet. Now it is a different story if you have no teeth to chew—well, if you have no teeth, then you can use small amounts of liquid frequently—but that is a special therapeutic case. Under normal circumstances, for normal people, who have some teeth yet in spite of so much sugar in the prevailing diets, there is no necessity to eat things in the form of juice. You can eat a carrot grated quickly, fresh, mixed in your vegetable salad—not eight carrots in the form of juice, but one carrot. It will be fine. And you can put a tablespoon of nice, tender, fresh green grass like this into your vegetable salad or other dishes as it is—it will be a nice tender fiber containing chlorophyll and good vitamins—and your body will be able to utilize it fully. Very often, simple common sense will tell you what you need, without a knowledge of biochemistry.

*What do you think of the motto, "whole foods for whole people?" Should foods be eaten whole, and not fragmented?*

Well, that's a very fine principle because when you make juice you lose the fiber and other important components of the fruit or vegetable. Of course, we cannot carry the principle dogmatically or rigidly—suppose you eat a whole coconut—you will be in the hospital tomorrow. We must use common sense. And apropos, I want to give you some good advice. You are here now in an unusual place—you are not used to all these exotic fruits and foods that you find in the markets—don't try to eat them whole! I recommend you always wash them very well, then always peel them, and eat only the inside. But stop at the hard seeds, because many of these delicious and juicy fruits have very hard seeds. There is a Latin expression: *est modus in rebus*—there is a way in all things. You may have a very serious stomach upheaval if you buy fruits in the market and try to eat the skin and seeds as well as the inside. There is a wonderful Spanish proverb which says, "es mejor tener miedo que asustarse." It is better to be afraid (take precautions) than to be frightened.

*In sprouting seeds, can you use cotton instead of a glass jar?*

The glass jar is the easiest way. And there is one thing which can happen with cotton: very often it develops mildew, while with this method there is no mildew because there is always air—you must always have air—if you prevent the air from reaching the sprouting seeds you will get mildew, and that is not so good. In the glass jar it is easy to rinse the sprouts morning and evening, so there is always moisture, and they are continuously exposed to the air—it is the simplest, fastest and most efficient way.

*Can people get along with only moisture and not water, too, like plants?*

Well, I don't recommend it. I don't see any reason why, when you are thirsty, you shouldn't drink. I remember good old Count Leo Tolstoy—he was the greatest novelist of all times, but he had some funny ideas—in many ways he was revolutionary, but in others he was very conservative. I don't think he would be very popular with the Womens' Lib movement, but once he said, "for a woman, it is enough to know to go under the eaves when it rains, to drink when she is thirsty, and to eat when she is hungry." I cannot resist my sense of humor to quote him. Now here again, you remember that by breathing you lose a lot of water from the organism—then also you are breathing not only through the lungs, but also through the skin—you have no idea how much invisible evaporation of water is taking place through the skin. You are breathing twelve thousand quarts of air a day, and with every breath you are losing moisture—therefore, it must be replaced. When you feel thirsty, it is simply the voice of the organism telling you to drink some water. It is very important—I don't recommend you try to live without water. When you asked if we could live without it, it reminded me of a gentleman in Germany in the 19th century called Schrott, who invented the Schrott-cure. He had an ingenious idea—to dry out the organism from all supposedly harmful liquids and humors, etc. Of course, he didn't know much about biochemistry—he let his patients eat dry bread several times a day, and reduced the liquid to the extreme minimum. Imagine what was that liquid—a little wine—one ounce of wine at a time, and a little dry bread. Well, this was an extremely heroic diet—most of his patients ran away on the third day. But some of them persevered, and the humorous thing was that he was able to cure syphilis with that diet! He became very famous for that cure, because those who were able to survive the diet got cured of syphilis. But please don't do any funny things like that because it is not necessary—why shall we kill a fly with a sledgehammer? It is not a good idea to dry out the organism.

*Is there any way to drink the dew?*

If you read volume one of my memoirs,* you will know that drinking the dew once saved my life. It was in 1938 that I took a boat from Chetumal, in Mexico, hoping to reach Tampico—but the boat disintegrated in a storm—the crew forgot to listen to the weather reports—and only my experience in long-distance swimming enabled me to reach a small deserted island which definitely was not the paradise people think of when they imagine remote, romantic desert islands! It had no water at all, and more rocks than trees. I felt it was a somewhat inglorious way to die, and I still had many books I wanted to write—and also, the ship's captain and cook had also managed to swim to the island, and they were looking to me to find a way out of our dilemma. So I was meditating, and suddenly I remembered Plinius, and his description of the Essenes, and how they utilized the heavy dew of the morning, not only to grow their plants and trees, but also to provide themselves with the moisture they needed to live. He used the words, "they partake of the morning dew from the desert flowers and plants," and went on to describe how they spent about half an hour each morning licking the dew from those plants. Well, I looked around and realized that the preconditions for a heavy morning dew were similar to the desert, and although it was very hard work and took a long time, by going from plant to plant on all fours, licking the dew, we managed to survive until we were rescued. I don't want to go into details, because I wrote

*Search for the Ageless, Volume One: *My Unusual Adventures on the Five Continents in Search for the Ageless.*

about it in my book, but thanks to Plinius, I discovered through the empirical method that you can, in an emergency, fill the body's requirements for water by drinking dew. But I don't recommend becoming shipwrecked on a desert island just to try it!

Well, I think we better return to our basic material and the rest of the questions I will take care of tomorrow. Before we talk about the uses of the Portable Meadow and the Biogenic Battery, I want to explain to you a few very important principles. You see, in the human body we have an organovegetative system and we have a cerebrospinal system. The organovegetative system is the seat of all our automatic involuntary activities—well, for instance, breathing, and all kinds of other basic biological functions which are going on without our conscious participation. The human body is a tremendous biochemist—it knows exactly how many phagocytes or leucocytes or hemoglobins or certain hormones or enzymes we shall create, what we need. We couldn't calculate these complex things with the most advanced computers, and these automatic processes are going on continuously in the body. Our organism is an extremely complex laboratory, which nature did create through millions of years of phylogenetic heredity, and fortunately it is going on without our participation—because we surely would make a mess of it if we were to interfere with these basic biological functions. In our genes, in our DNA, etc., we did inherit millions of years of phylogenetic tendencies from our long line of ancestors, and these are certain cosmovital forces which in our planet are working in every living organism for millions of years. These functions are perfect because there are millions of years of wisdom behind them. This is why I say it is very fortunate that we cannot monkey with them, or disturb them, or implant funny ideas asking these basic functions to stop. For instance, suppose you read some strange book that says it will be wonderful if we stop breathing—well, don't worry, you will continue to breathe. If you think, oh, we could reach eternal life if we could stop the circulation of the blood, well, don't worry, it will go on. Absolutely no funny ideas can interfere with it. This is a kind of primeval power in every living organism which is going on without our understanding and without our knowledge. And we can always depend on that. It is the accumulation of millions of years of the wisdom of nature.

Now human intelligence is only some thousands of years old.

The function of the cerebrospinal system did create something above nature, and these are our voluntary actions—our thoughts, etc. Now there is nothing wrong with that—we shall not underestimate it. Our cerebral functions did create the masterpieces of Michelangelo and Leonardo da Vinci and Bach, Mozart, Beethoven, and Shakespeare and Einstein, etc., and all that is fine, we have nothing against it—in fact, we shall try to perfect it, because as we are now, we don't use more than 85% of our brain cells—there is a tremendous possibility to develop our capacity, to improve this mathematical proportion. But our main purpose should be that our conscious cerebrospinal functions should be in harmony with the primeval basic power of life which governs our organovegetative system and the whole life of our planet. Our troubles and problems appear when we deviate from this primeval force which is sustaining us, and then we become at war with ourselves. In one of my books, *Cosmos, Man and Society,* there is a chapter: *Man at War with Himself.* Whenever we have disharmonious thoughts or emotions, we interfere with this basic, primeval, tremendous and vital force which is carrying on in the organism—this magnificent biochemist and scientist—this profound and mysterious power which took nature millions of years to perfect. When we have the wrong diet of thoughts and emotions we create nervous tension, we become tired, we create unbalanced disharmonious secretions of hormones and enzymes—we interfere with this basic power of life in the universe, and especially on our earth. Now the science of Biogenics shows you how you can train your cerebrospinal system to function in harmony with this primeval force of life which created us as we are, through the phylogenetic work of millions of years. Because if our conscious activities will be against this primeval biogenic force, in that case we will have all kinds of troubles, as you can see everywhere you look in the twentieth century. There never were in history so many people with nervous tension, nervous breakdowns, all kinds of emotional troubles and ailments.

I want Norma to tell you about certain basic biogenic techniques, and a little later I will talk about Biogenic Meditation, which will carry us into direct contact with this primeval life force—this biogenic power which creates certain physiological and biochemical changes in our organism.

*Note from Norma: The best explanation for the reader is to reprint here those pages from our I.B.S. guidebook,* The Essene Way-Biogenic Living, *which deal with these subjects.*

*Fifty million years ago, our guardian angel, the grass, arrived to our planet—to make life possible, and to prepare the earth for the human race. . .*

### THE MIRACLE OF GRASS

In order to understand the principles and purpose of Biogenic Living, we must first understand the importance and significance in all our lives, of that most humble, ubiquitous and universal of all plants—the grass. According to Sir Fred Hoyle, "the emergence of intelligence on Earth was probably due to a combination of several circumstances, among which the most important was the development about fifty million years ago of the plant now called grass. The emergence of this plant caused a drastic reorganization of the whole animal world, due to the peculiarity that grass can be cropped to ground level, in distinction from all other plants. As the grasslands spread over the Earth, those animals that could take advantage of this peculiarity survived and developed. Other animals declined or became extinct. It seems to have been in this major reshuffle that intelligence was able to gain its first footing on our planet."

The climato-meteorological consequences of the gradual destruction of the rain forests of the world are resulting in devastating dry spells in many parts of our planet. When you live in an environment of concrete, metals, plastic, and other artificial substances, losing your contact with the elements of nature, and mainly with their ancestral synthesis, the plant world, your biopsychological condition will gradually deteriorate. When, in addition, you breathe polluted air, and your body has no more contact with green plants, etc., your individual biosphere gradually diminishes and deteriorates. The generally accepted psychoneurotic problems of the city dweller stem from this destruction of the individual biosphere. The process seems to be not only inevitable but insurmountable.

While you are not able to improve the biosphere of the whole planet, you can definitely ameliorate at least your immediate micro-environment, in your own room, for example. When your vitality inexorably diminishes because of the absence around you of fast-growing, life-generating, air-filtering young plants, you can do something to reestablish your lost symbiosis with vital, young, fast-growing plants and the more natural air created by them in your room. Remember, you are breathing twelve thousand quarts of air daily. It is our most important food.

It is possible that the phylogenetic longing for contact with the vital, green biogenic zone of our planet, imprinted on the DNA of our cells for hundreds and thousands of years, is the reason for the recent popularity of house plants in the homes and apartments of city dwellers. But no nursery-produced house plant can equal the power and biogenic vitality of the young, fast-growing grass in our *Portable Meadow* and *Biogenic Battery*. With very little effort and expense, we can actually create a mini-forest within our own dwelling, perhaps not as ideal a setting as the cathedral-like forests where our distant ancestors lived in perfect symbiosis with the biogenic forces of nature—but the reality is that we live here and now, in the midst of a highly technological era at the latter part of the twentieth century, and we must do our very best with the tools we have. And with these simple tools—the Portable Meadow and the Biogenic Battery, we can create a self-perpetuating, biogenic field of vital forces to surround us, and we can enjoy the same phylogenetic symbiosis with the primeval life forces which our ancestors enjoyed for hundreds and thousands of years.

### HOW TO MAKE A PORTABLE MEADOW

*"Ours is an unswerving 'Custer's Last Stand'*
*Against the flood of concrete, metals and plastics*
*And the pollution of air, water, soil, and vegetation—*
*That man may live."*

Begin with a simple plastic paint bucket from any hardware store, about 5 to 6 inches deep, and about 10-12 inches in diameter. Fill it with good soil. If you have an outdoor source that is free of pesticides and is of good porous consistency, use it. If not, use an organic potting soil from any nursery, large hardware store, or even a supermarket.

Soak one-half cup of whole wheat grains (or most other whole grains) in water of room temperature in an open glass jar overnight. The next morning, put a layer of cheesecloth over the jar opening, fasten with a rubber band, and rinse the wheat thoroughly by letting water from a faucet (tepid temperature) run into the jar, then turning it upside down and letting the water run out. Repeat this rinsing several times until the water is clear.

Set the jar with the rinsed wheat grains in a diagonal position (a dish drainer is excellent for holding it in a diagonal position) in a dark place for 12 hours. Do not let anything block the cheesecloth-covered

In just seven days, a handful of wheat grains becomes a Portable Meadow—a lush, dense mini-forest overflowing with primeval biogenic vitality. *(below)* Not only wheat, but lentils, onions, garlic, etc. may be planted in simple buckets, giving a constant supply of tender, delicious biogenic baby greens.

opening, as the wheat grains need air. Also, if a dark place is unavailable, simply cover the jar lightly with a porous towel.

After 12 hours, rinse the wheat again, and let it sprout for another 12 hours. After a full 24 hours of sprouting, rinse the grains thoroughly; as described above.

Now spread the sprouted wheat grains like a carpet over the top of the soil in the bucket. Wet the grains thoroughly, but do not completely saturate the soil with water, as there is no drain hole in the bottom of the bucket.

On one side of the wheat grain is the beginning root; on the other end will come the sprout, which eventually will become grass. It takes about a day for the little roots to find their way into the soil, and until they do, it will be a good idea to keep the grain-carpet moist and out of direct sunlight. If you work and have to be away all day, cover the bucket with a large plastic bag, poking a hole in the top for air. This will keep the moisture in, but do not forget to remove it after 24 hours, or the grains may start to mold.

As soon as you see the white shoots start to come up, you can place the bucket in a window-sill, or anywhere the grains will receive light (it does not have to be direct sunlight). Only an occasional moistening now will be needed, as the shoots very quickly turn green, and the growth from now on is amazingly rapid.

In seven days, the young biogenic grass will be ready for those biogenic practices described in this book.

### HOW TO MAKE A BIOGENIC BATTERY

The biogenic battery differs from the portable meadow only in size. It will be a good idea, before starting to plant the first biogenic battery, to acquire about two dozen small cup-size containers in which the opening is about the same diameter as the base: these can be yogurt cartons, 8 oz. cottage cheese cartons, cardboard cup-size containers from a nursery, even glass or plastic wide-mouth jars that held cosmetics or shampoo (but be sure to wash them very well.)

Start the grains soaking as for the portable meadow, but of course the amount is greatly reduced. Do not soak more than one flat tablespoon of grains at a time. Soak the grains in the same way, in an open glass jar, or drinking glass, with the opening covered with one small layer of cheesecloth fastened with a rubber band. The next morning, let faucet water run into the jar, then out again (the cheesecloth will prevent the grains from escaping) and once or twice more to be sure they are thoroughly rinsed. Then set the jar or glass in a diagonal

*The Biogenic Battery*

position (it is so small, it will fit on your dish drainer right there in the kitchen) for 12 hours, covering the jar with a small piece of paper towel to keep the grains in darkness.

After 12 hours, fill the cup-size container with soil (not quite to the top) and spread the wheat grains (which by now have started to sprout) evenly over the top of the soil. Moisten the grains thoroughly, but be very careful not to add too much water—remember you are dealing with a very tiny amount of soil.

Cover the little container completely with a small plastic bag (you can poke a tiny hole somewhere in it for air), place a piece of paper towel over that, and set it aside for 12 hours. The plastic bag assures that the grains will stay moist while the little root-sprouts are finding their way into the soil, and it also makes it very convenient for those who have to work all day and cannot be home to occasionally moisten the grains.

After 12 hours, moisten the grains again and replace the plastic bag. In another 12 hours (24 hours in all using the plastic bag) remove the plastic, and place the little container on a window-sill. Only very occasional moistening will be necessary now as the white shoots will come up very rapidly, turn green, and soon be transformed into tender green leaves.

After seven days, and until the fourteenth day, the biogenic battery will be ready for the biogenic meditation. From the fourteenth until the twenty-first day, the biogenic battery will be ready for the biogenic fulfillment practices.

### CONTINUOUS PLANTING OF THE BIOGENIC BATTERY

Obviously, you will have to have a system of continuous planting—otherwise, you will be able to take advantage of the multiple uses of the biogenic battery only once a month. A 7-day cycle is the best, and once the routine is established, you will find it is extremely easy and becomes almost second-nature.

Before we continue with a description of some of the uses of the Portable Meadow and the Biogenic Battery, there is an anticipated question I want to answer, for those who wonder why they cannot simply go outdoors to their lawn, their flower garden or vegetable patch, and receive the same benefits of the biogenic force field? Why is it necessary to plant whole grains in small buckets and containers when outdoors (if you are lucky enough not to live in the city) there are all kinds of plants, trees and grass growing? The answer lies in the word *biogenic.* When you plant these whole grains, within a short seven days the grass is several inches high—a veritable starburst of explosive, vital, life-generating *biogenic* energy—energy which the fast-growing plant shares with you—energy which surrounds you, flows through you, nourishes you, and revitalizes you. Once the plant reaches its *bioactive* stage, after about three weeks, the plant still has plenty of energy to maintain itself, but not enough to share with you. When it gets older, and reaches its *biostatic,* or aging phase, then it has just enough energy to barely maintain itself. And of course, the *biocidic,* or dying, state follows naturally, as the plant returns to the earth which gave it life, and a new cycle begins. The grass on your front lawn and the vegetables in your garden are most likely in the bioactive stage and have no excess energy to share with you. This is why we recommend, as the first step on the road to Biogenic Living, the planting of the Portable Meadows and Biogenic Batteries, to provide you with a continuous supply of vital, life-generating biogenic energy, from the vast primeval lifestream of our planet, of which we are an inextricable part.

In addition to providing biogenic energy simply by sharing our existence, there are many things we can do with the Portable Meadow and Biogenic Battery to increase our health, vitality, and even our individual evolution. In following chapters we will learn some of the most important and ancient uses, and here I would like to describe some of the more simple ones, extremely important to our survival in this technologically-oriented century, when our air, water and food are so vulnerable to contamination. By the correct use of the Portable Meadow and the Biogenic Battery, we can greatly improve the quality of our air, our sleep, our ability to relax, and even our personal hygiene.

### BIOGENIC RELAXATION IN THE MEADOW

I do not need to impress on the reader the vital need for a healthy and efficient method of relaxation in our tense and troubled world. Tranquilizers and sleeping pills are dispensed like candy by physicians who should know better, but even these harmful chemical crutches do not help to alleviate the underlying anxiety which accompanies our steady drift away from our phylogenetic world of green grass and tall trees.

Unlike artificial and harmful methods, this method of biogenic relaxation *works*—because it utilizes the powerful biogenic force field in the Portable Meadow. The silence which surrounds you is the silence of the forest—alive with growth, vitality, and the quiet joy of nature.

*The method:*

1. Place the Portable Meadow on a table.

2. Bring a comfortable chair to the table.
3. Sit in a *comfortable* position in the chair for ten minutes, as close as possible to the Portable Meadow.

The biogenic force field emanating from the meadow surrounds you and penetrates into your whole body for ten refreshing minutes, as you absorb it. You can enjoy this biogenic relaxation in the meadow whenever you feel tense or tired.

The ancient Essenes knew that sleep, in addition to providing the body with the preconditions for biological repair, can be a source of the deepest knowledge. They believed that when the last thoughts before sleep were harmonious ones, the subconscious would be put in contact with the great storehouse of superior cosmic forces, described in greater detail in the chapter *Biogenic Psychology*. But it is often extremely difficult these days for people to achieve a really good night's sleep. Digestion may be working overtime, unsolved problems and anxieties may be fermenting in the mind, and in the city, pollution and absence of growing green plants may deprive the brain of the oxygen it needs for restful sleep.

By utilizing the powerful biogenic force field of the Portable Meadow through the night, the quality of sleep is vastly improved. Not only is the oxygen content of the air increased, but the vital, biogenic energies absorbed contribute to our gradual ability to contact more and more the regenerating flow from the primeval Ocean of Life.

*The method:*
1. Place the Portable Meadow on a small, eighteen-inch tall table, as close as possible to the head end of your bed.
2. Just go to sleep at night as usual, making sure you have some fresh air.

The biogenic force field will surround you and penetrate into your whole body during the night, while you absorb it.

BIOGENIC BREATHING IN THE MEADOW

Complicated breathing exercises are completely superfluous; it matters not *how* we breathe, but *what* we breathe. Brisk walking in fresh, pure air is the best breathing exercise; complex techniques of breathing performed in a stuffy room in the middle of a polluted city can do only harm to the body. Air is our most important food—we take in more than twelve thousand quarts of it a day. No matter where we live, by adopting this simple method of biogenic breathing in the meadow, we can create a vital, oxygen-rich atmosphere—even while driving in our cars (the Portable Meadow is much more efficient than air conditioning to provide a biogenic atmosphere in the auto, even in the midst of heavy traffic with its choking fumes).

*The method:*

1. Place the Portable Meadow on a table.
2. Bring a comfortable chair to the table.
3. Sit in a comfortable position in the chair and lean forward toward the table, supporting your elbows. Then inhale and exhale gently, without strain, your breathing directed toward the meadow. Breathe in and out altogether seven times, with short, natural intervals between exhaling and inhaling.

You breathe in the subtle, new-mown haylike fragrance from the green leaves of grass, with the emanating oxygen, and exhale the carbon dioxide waste from your lungs. This sevenfold biogenic breathing can be enjoyed at any time you need refreshment. (To use while driving, simply place the Portable Meadow on the seat next to you and forget about it. It will do its biogenic work while you keep your eyes on the road!)

### BIOGENIC DEW BATH

The biogenic dew bath is an adaptation of a very ancient technique used in many civilizations of the past. It utilizes the refreshing dew from tender young grass in order to cleanse, invigorate and revitalize the whole body. It is also an ingenious way to thoroughly wash the body using only half a pint of water! That is quite a contrast to the gallons and gallons wasted in a traditional shower or bath. The biogenic dew bath is taken with the Biogenic Battery, not the Portable Meadow.

*The method:*

1. Undress in a room or outdoors in a comfortable temperature.
2. Dip the grass of the Biogenic Battery (holding the container firmly) into water of room temperature. Be sure not to get the soil wet, only the grass (which should be at least 10 days old).
3. Gently and quickly sponge all accessible parts of the body with the wet grass.
4. After your short dew bath, walk around for a few minutes until you are dry.

The fine capillary nerves and veins in your skin will be stimulated and refreshed, transmitting this fresh and invigorating feeling through the whole body, parallel with the absorption of biogenic energy from the force field of the Biogenic Battery. You may enjoy this biogenic dew bath any time you feel the need of refreshing invigoration.

One important thing to remember: the container, seeds and soil used for the Biogenic Battery must be very *clean* and untreated with chemicals to prevent any skin trouble through contact.

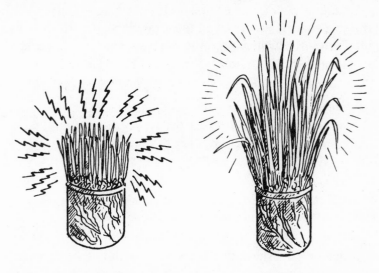

*(Professor continues)*

All these uses of the Portable Meadow and the Biogenic Battery will put you in contact with the biogenic forcefield which surrounds the fast-growing, tender greens. We can be very grateful to Kirlian photography, because with that technique for the first time we were able to measure and represent the field of forces around the biogenic grass. All you have to do is go close to its sphere and it will impart to you that exuberant excess of biogenic power—this is the field of force which is creating that relaxation and the other effects from the different uses of the Portable Meadow and the Biogenic Battery.

I want you to realize something: if you were to represent the history of our planet as a tape a mile long, the last sixteenth of an inch would represent the known history of mankind. Therefore, we shall be conscious that we have in our genes, in our DNA, in our heredity, tremendous capacities and forces on the biological level which we did inherit from our ancestors millions of years ago—even before our ancestors became humans, not only since *homo sapiens* has existed. Therefore, when you use the Portable Meadow or the Biogenic Battery in one of these applications, the field of force surrounding you, shown by Kirlian photography, has a multiple effect—not only the oxygen-forming effect, which you breathe in, but many other effects on your organovegetative system—and that affinity, that millions of years of affinity appears right there, and you become one with this biogenic field of force, and it disconnects you from all the erratic activities of the cerebrospinal system which may cause nervous tension, disharmonious thoughts and emotions, or interfere with your sleep.

*What is the area of the force field which surrounds the plant?*

It depends on the age of the grass. At this age of five days, for instance, you may consider that the area of the force field is more or less two or three feet. The closer you go to the plant, the more intensive it is. Kirlian photography shows a much denser, stronger force field closer to the plant, which weakens as it goes outward, and at about a distance of three feet, disappears.

*Did the ancient Zarathustrians use the Dew Bath?*

Yes, and they used the dew itself. But it will be difficult for you to gather dew in the city—therefore, the dew bath is excellent.

Perhaps the most important use of the Biogenic Battery is in the practice of Biogenic Meditation. When you meditate in this way, you don't have to imagine anything, you don't have to think of anything or not think of anything—just simply acknowledge the reality of this primeval power—you put yourself, your whole body in the field of force of the biogenic battery, and absorb all the exuberant, excess vitality from the fast-growing little baby plant. That force field represents a primeval power built and perfected by nature through millions of years, and you are absorbing that excess cosmovital energy, that biogenic power, from the Biogenic Battery. This, besides disconnecting you from the erratic activities of the cerebrospinal system, connects you immediately to the most powerful source of energy in existence—that same biogenic primeval power—and this brings also profound biological changes in the organism. I want Norma to read that page in *The Essene Way—Biogenic Living* which describes these physiological effects. Now I don't speak about the main effect, about uniting us with that primeval power and disconnecting us from the source of disharmonies which we develop in our way of conscious thinking, but I speak about scientifically measurable products and consequences in our biology, physiology, and biochemistry.

*(Norma reads)*

When the young, biogenic plant reaches the seventh day, the fast-growing, deep green robust grass is about six inches tall, its roots completely filling the whole container in a tumescent, hard condition. Your biogenic battery is now ready for use.

To perform the biogenic meditation, place the plant on a table, about one foot from the edge. Sit easily in a comfortable chair, facing your biogenic battery. After a minute or two of relaxation, grasp with both hands, between your fingers, as in prayer, the young biogenic grass plant, close your eyes, and hold them for about twenty minutes (elbows resting comfortably on the table), feeling between your fingers the living, dewy freshness of the young plant, and feeling the life-forces entering into your whole body, through tactile contact with your biogenic battery. As it is very important to be comfortable for these twenty minutes, it will be a good idea to support your elbows on something soft (a quadrupled soft towel, for example). Also, let your palms rest easily on the rim of the container. If you want to think on a word at each breath, you shall use only the word "Life," because *it is, and it corresponds with the reality.* You will feel the surplus biogenic, life-generating forces from the vigorously-living young plant flowing continuously into your whole body; soon after this flowing sensation, you will feel a tingling sensation, especially through your spinal cord, and finally, this powerful lifestream will shake your whole body from time to time. After approximately twenty minutes, open your eyes, remove your palms and fingers from the plant, rest for a few minutes, and, refreshed, proceed with your daily activities. You shall repeat the whole procedure once more in the evening, preferably sometime before dinner (never after meals) and not just before going to bed, as the powerful biogenic energy absorbed may keep you awake for several hours.

This biogenic meditation will free you twice a day from the tyranny of your restless, tense, worrying cerebrospinal functions (the seat of your voluntary processes) and will keep you in contact exclusively with your organovegetative system—millions of years older and incomparably more powerful—working independently from your conscious, voluntary actions and mistakes. This primeval, infallible supercomputer which performs every second for you the most complicated mathematical and biochemical calculations,

*Professor illustrates the position of the hands during Biogenic Meditation.*

knowing exactly how many millions of different cells (leucocytes, erythrocytes, phagocytes, enzymes, hormones, etc.) to synthesize in the thousands of cellular systems of your organism, is the omnipotent and omniscient Law of Life, directing all manifestations of Life on our planet.

Actually, this biogenic meditation is really not just meditation, but in reality a powerful biodynamic union with the greatest primeval power on earth. As Kabir, the great poet-saint of ancient India, said:

> . . . It has no end, nothing stands in its way.
> Where the rhythm of the world rises and falls,
> thither my heart has reached. . .

There are hundreds of different meditation methods in vogue, but these only temporarily free you from certain tensions and stresses, "connecting" you mostly with imaginary, unreal hypotheses, but not connecting you in a real, tactile way with such a *real* and tremendous power as this *vital, life-generating, primeval lifestream.* While you are practicing the biogenic meditation only twice a day, the cumulative repetitive effect is constant, and affects your behavior twenty-four hours a day. You will soon discover that you feel refreshed, vital, full of energy, finding an inner peace and harmony with nature, society, and culture. You will experience a complete regeneration in feeling, thinking, and acting, achieved by the security that you have permanent access to all the sources of harmony, energy and knowledge. You may not know everything—but you will know all things which are necessary for your happiness, and that is not an easy achievement. Happiness is difficult to find within ourselves, and impossible to find elsewhere.

> "The heavens smile, the earth celebrates,
> the morning stars sing together,
> and all the Children of Light shout for Joy."
> —The Essene Gospel of Peace,
> Books II and III

> "The moon shines in my body, but
> my blind eyes cannot see it:
> The moon is within me, and so is
> the sun.
> The unstruck drum of Eternity is
> sounded within me,
> But my deaf ears cannot hear it. . .
> —Kabir

As we go farther and farther from our natural, primeval biogenic state and environment, with more and more concrete, plastic and metal, and less and less meadows and greenery at the end of the twentieth century, the aberrations of our cerebrospinal functions are creating inexorably a progressively biocidic environment in alarming proportions, polluting the air, water, soil, vegetation, creating even light and sound pollution, and proliferating radiation fallout and thermonuclear arms. We ourselves are becoming victims of the stress, tension, worry, confusion, and pathological states of the mind, demonstrating the fatal example of the results of our cerebrospinal violations of and inharmony with our fundamental organovegetative functions, inexorably and phylogenetically united with the primeval ocean of life in and around us.

The biogenic meditation in oneness with the cosmovital forces emanating from the tactile contact with the biogenic battery, not only brings our erratic cerebrospinal system into greater and greater harmony, through cumulative effect, with our inner and outer biogenic environment and functions, but also proves the efficiency of these biogenic meditations (communions), by producing measurable improvements in our psychophysiological states, more and more similar to biogenic plant functions. This beneficial switch from our tense cerebrospinal functions toward the peaceful, relaxed organovegetative hypometabolic state, produces palpable, plausible, and scientifically and quantitatively measurable facts:

1. During these biogenic communions, there is a considerable lowering of oxygen consumption (an average of from 250 to 210 cubic centimeters per minute, much closer to plant metabolism, enabling us to exist with less oxygen intake (which may become in an emergency extremely important).

2. We found also that carbon dioxide elimination was lowered (also much closer to biogenic plant metabolism) from 220 to 185 cubic centimeters per minute. This, in case of an emergency of being trapped in a closed space, also could prove very useful.

3. Both the rate of respiration (from 16 breaths per minute to approximately 9 per minute) and the volume of respiration were considerably reduced, representing a greater oxygen economy, closer to plant-like behavior.

4. Our stress-created waste products, like blood lactate, etc., are very considerably decreased, little by little approximating the hypometabolic state of the relaxed, growing young grass.

5. High arterial pressure starts to gradually decrease, reaching a normal level, further indication of complete elimination of stress.

6. The heartbeat and pulse-rate become more and more normal, indicating a much easier, more effortless function of the heart.

7. Sensibility to pain is markedly lessened, another organovegetative plant quality.

In addition to the above scientifically measurable and verifiable facts, against which arguments have no value, we may add several mental achievements, outside of the duration of these biogenic meditations (communions), also similar to biogenic plant states and conditions. For example:

1. A state of relaxed alertness
2. Improved concentration
3. More adequate response to stressful conditions
4. Faster and more adequate reaction time
5. Increasing, then complete relief from insomnia
6. Cumulatively less anxiety
7. Gradual improvement of memory

All the above improvements created by the continuous, uninterrupted daily biogenic meditations (dynamic communions with life-generating forces) amply indicate that our overworked, erratic, tense cerebrospinal functions are transformed into ever-increasing harmony with our primeval, phylogenetic organovegetative system, and the ancestral primeval affinity is reestablished. This potentially existing phylogenetic affinity makes it possible for us to achieve the above outlined improvements with only two daily twenty minute meditation-communions, but with a constant, twenty-four-hour-a-day lasting effect.

*(Professor continues)*

In that biogenic state, the organism and the heart can function more perfectly, and much less oxygen is needed—this ability could in some critical condition save our lives.

Remember, please, the words "closer to plant life." For that short period of twenty minutes or so, we achieve that unity with the primal organovegetative state, and we are closer to plant life. Please remember that this little baby plant of five days old is really fifty million years old—it has the phylogenetic and biological heredity of fifty million years. This is why it is so powerful. If you look around the world, you will see grass everywhere. In the hottest Central Africa, in the coldest part of northern Siberia, you will find it in every climate and in every geographical area. When heavy machinery goes through a place, well, it squashes completely the earth—next day, you will find grass growing again. When something happens to break a concrete sidewalk which has been there for fifty years, immediately grass starts to grow in the cracks. It has a tremendous inherited vitality of fifty million years—it can grow anywhere and everywhere, without cultivation—it is the most perseverant, the strongest, the most tenacious living being on our planet. I will say again how grateful I am for Kirlian photography which came to verify my thesis expounded in 1928 in my book, *La Vie Biogenique*—it proved the existence of the force field of biogenic energy surrounding the fast-growing grass, a force field we can dip into for a living, eternal and real source of constant, vital power. It is logical that when we need a source of energy, when we are run-down, or tired, or tense, that we should draw it from the original and most powerful source, which is the recently-born tender, young grass, with all its primeval force.

There is a beautiful legend in Greek mythology I want you to know. Anteus, the son of the goddess of the earth, Gea, was in battle with Hercules, the strongest man on earth. Being so strong, Hercules over and over threw Anteus to the earth. But each time Anteus fell to the earth, he received a tremendous surge of energy from his mother, Gea, the goddess of the earth, and, renewed with strength, jumped up again to fight. Each time Hercules threw him down, he was getting from his mother, the earth, renewed force and vitality. And this is what is happening to us. When we practice biogenic meditation, we instantly come into contact with fifty million years of accumulated power—we become practically one with this young, exuberantly-growing grass. Our metabolism descends to the quiet, radiant energy of plant level, where everything is harmonious, everything is relaxed, and the tremendous sources of energy from the field of force around the biogenic battery flow through us. So always remember Anteus, whenever you have a serious crisis in life, you run into all kinds of troubles, you are insecure, you are worrying—get strength from the goddess of the earth—get strength from this fifty-million-year-old source of energy—for twenty minutes here and there, enter your primeval, ancestral home of oneness with the biogenic lifestream.

*Professor, what relationship does the biogenic meditation have with the sixteen forces of Zarathustra?*

There is not only a relationship between them, but I may say that this biogenic battery is a microcosmos, an embodiment of all the sixteen forces.

Lao Tzu said that one picture is worth a thousand words. I want Norma to read to you the most important chapter in *The Essene Way—Biogenic Living,* called *The All-Sided Microcosmos of the Biogenic Battery,* and at the same time place one of the biogenic batteries on the symbol of each of the sixteen natural and cosmic forces. And while she is reading, I want you to meditate on the profound meaning of each of the forces, and how, through direct contact with the biogenic battery, you can draw to you their wisdom, energy and harmony.

*(Norma reads)*

For the benefit of those who may consider, even after all the material in this book, that the biogenic battery is, after all, just a cup-size container of soil filled with growing wheat grass, I would like to describe, in simple terms, the immense and profound significance of that miniature biogenic forest we can hold in our hands. Just as Zarathustra was able to find the universe in a grain of wheat,* so can we discover every one of the sixteen natural and cosmic forces in the biogenic battery.

This idea is not new. In the manuscript found at Monte Cassino, Plinius, the Roman natural historian, described how the Essenes always had a few benches in their dwellings where they kept small earthenware pots, and in these pots were growing ordinary herbs and grass. These fast-growing herbs were not used only for eating, according to Plinius. The Essenes considered that the dense mini-forests they cultivated in earthenware pots were a representation of the whole cosmic order, and they used them in their communions with the angels to remind them of the totality of the Essene teachings and traditions.

*Sun* is represented in the biogenic battery through the green color of the leaves of grass. The green is chlorophyll, produced by photosynthesis, made possible only by the presence of sunlight. We see the power of the sun through the beautiful green color of the living grass.

Just as there would be no organic life on our planet without the sun, so would all life perish without *Water.* In this exuberant

*The beautiful legend of Zarathustra and the grain of wheat is told in *The Essene Book of Asha.*

161

Professor describes in his inimitable way the multiple uses of the Biogenic Battery, and the fact that in the five-day-old baby plant there is really fifty million years of accumulated biogenic power—a forcefield of biogenic energy we can learn to absorb and utilize to improve every facet of our lives.

growth of tender young greens, there is more than seventy per cent of that wonderful water of life of the Essenes. When we touch the grass with our hands we can feel the dewy freshness of the water-blessed living grass.

Regarding *Air,* the biogenic battery is like a factory of intensive manufacture of oxygen. Particularly in its biogenic stage, it is producing constantly the purest form of air for our benefit, continuously cleansing and refreshing our immediate atmosphere.

As far as *Food* is concerned, the biogenic battery *is* food—of the purest, most nutritious kind. When the leaves are only three inches high, we can cut them and mix them with a vegetable salad, a cereal, or any other wholesome dish. The tender leaves of grass are bursting with vitamins, minerals, enzymes, and plant hormones.

*Earth* is also right here in the biogenic battery, for it is earth which provides the nourishment so the leaves of grass can grow. This little handful of soil contains all the elements, all the material preconditions for the growth of this green grass. Before our eyes we watch the miracle of birth as the wheat grain swells, the roots start their journey downward to draw life from the soil, and the sprout bursts open with a tender shoot of vivid green, bravely overcoming gravity in its upward ascent.

*Health* is nothing else but the dynamic unity of all the other earthly forces. There is nothing healthier than young, fast-growing grass—it is a true representation of health, as it unites all the forces of sun, water, air, food, earth, and joy.

*Joy* is a force one can experience through the practice of biogenic meditation. When we place our hands around the tender, young, living blades of grass, we feel the joyful exuberance of the primeval lifestream. Waves of joy surge through us and penetrate our bodies through tactile contact with the force field we share with the biogenic grass.

The first of the spiritual forces is *Power.* Kirlian photography has proved what the Essenes always knew intuitively—that fast-growing, biogenic grass has an intensive force field around it—tangible evidence of the life-generating power inherent in the living grass.

*Love* is another force which is experienced through tactile contact with the biogenic battery. When we hold the grass between our palms during biogenic meditation, a tremendous amount of excess energy flows to us from the biogenic battery. This gift to us of its energy is love—in its most direct and selfless form. The plant does not ask us if we are worthy or unworthy—it simply gives, unconditionally, of that magnificent energy which flows from the primeval lifestream of our planet through every living thing. There is a saying of ancient India: *only he gives, who gives himself.* This is what the little green plant is doing—it gives itself, its excess energy, to us. There can be no more beautiful example of love.

In the biogenic battery is the manifestation of infinite *Wisdom.* The coordination, the total preconditions, the utilization of all sources of energy, harmony, and knowledge are here in this living grass. As George Washington Carver said, "When I touch grass, I touch infinity." And it was Einstein who said, "He who does not feel awe in contemplation of leaves of grass is as good as dead; it is a manifestation of supreme wisdom." Goethe, perhaps, put it best of all: "Gray are all the theories, but green is the tree of life." This little biogenic battery is the Tree of Life.

*Preservation,* the foundation of ecology, is perfectly represented by the biogenic battery, for nothing is wasted in this little growing forest of grass. Everything is utilized, everything is absorbed, everything is preserved. Even in its final phase of life, the elements of the plant return to the soil and provide minerals for future generations of plants. It is a living example of the moving wheel of life and the teaching of Buddha that "nothing is lost in the universe."

Now we come to the most important force, that of the *Creator.* Could anything be more apparent? The biogenic battery is the creation of life before our eyes. Biogenic means life-generating, and here is the generation of life. We cannot help but be reminded of the Creator every time we see the miracle of creation take place in the biogenic battery.

*Eternal Life* is also evident in the biogenic battery, because it is the most evident characteristic of the humble grass, which defies the law of gravity by growing anywhere and everywhere, even between cracks in bricks, cement and concrete. Grass was here on our planet fifty million years before us, and it will probably be here for more than fifty million years after us. For life is not the exclusive privilege of our planet, which is a tiny point in our solar system, which is a tiny point in our galaxy, which is a tiny point in the ultra-galactic system, which is a tiny point in the known universe, and an even tinier point in the unknown universe. There-fore, it would be megalomania to think that life exists only on our planet. Life exists on billions of other constellations, for the universe is a mind-staggering infinity. And we are not the only children of the cosmos. We may say we are the grandchildren of the cosmos, and children of our planet. And between all the other cosmic grandchildren in the infinite cosmic space there is a solidarity of all forms of life. This solidarity between all forms of life on innumerable planets, according to Essene traditions, is the Cosmic Ocean of Eternal Life. Planets may disappear, solar systems may disappear, but the Cosmic Ocean of Life is eternal. And on this planet where we live, there is no more perfect representation of the Cosmic Ocean of Eternal Life than the fifty-million-year-old grass. When we touch grass, we truly touch infinity.

*Work,* creative work, is also a force very much in evidence. A seven-day-old biogenic battery is a wonderful example of intensive, creative work. We can practically see the roots in movement, the leaves growing before our eyes. A week ago this was just a few grains of wheat, and now we see the lush green evidence of intensive, creative work—a wonderful example for us in our lives.

*Peace* is yet another force which we experience during the biogenic meditation, as if the little plant will share its most important secret with us only if we put forth our best efforts. Through our diligent practice of biogenic meditation, little by little we are able to put out of our minds the "Grand Central Station" of the twentieth century, all the harsh invasion of technology and man-created cacophony. Then, at last, we are one with the Peace of Nature, with the great primeval peace of the biogenic lifestream.

There is one force I passed over in its proper place because I wanted to mention it last: the force of *Man,* the co-creator with God. In the beautiful Essene symbolism of the Tree of Life, man is shown in the center, half of his body connected to the earthly forces, and half connected with the cosmic forces—a meaningful illustration of the idea that man unites in his body and spirit all the forces of nature and cosmos in perfect harmony. When we practice biogenic meditation, we personify that symbol, for *we are here* in the center, physically here in the center of all these forces, our hands between the roots of the earthly forces and the grassy leaves of the cosmic forces. This is the true role of man, to be surrounded perpetually by the natural forces and the spiritual powers. In the words of Tolstoy, "Man is not alone in the universe; he is surrounded by infinite powers of love and wisdom." When we practice biogenic meditation, we are not just symbolically in the center of an imaginary tree of life—we are effectively here—we touch all these powers as only we, as human beings, have the right to do—and the Tree of Life becomes a tangible reality, a biogenic battery between our hands.

This is the meaning of the earthenware pots which the Essenes kept filled with fast-growing herbs and grass. As a representation of the totality of the laws of the universe, of all the natural and spiritual forces and our unity with them, it was a constant reminder for them to live like the Tree of Eternal Life. And so it should be for us.

### (Professor continues)

The reason I wrote this last chapter in the guidebook was to prove that in the biogenic battery all the sixteen forces of the Creation are represented, and this five-day-old baby plant has really fifty million years of accumulated biogenic power. There is no greater source of energy and vitality we could resort to, than this one. So please handle it with great respect. You may have some ancestors—your father, your grandfather, your great-great grandfather—but you don't have an ancestor as old and as powerful as this biogenic battery. You see, this is the real greatness in the smallness—a microcosmos, a miniature thing which embodies the whole cosmos, all the sixteen forces of the Creation, all in this little biogenic battery. All of us are the heirs of incredible wealth, because we did inherit this from the Essenes, and before that from Zarathustra, from ancient Sumeria.

### Why did we lose contact with these forces?

Well, you remember an interesting symbolic scripture—the Genesis of the Old Testament. Mankind from time immemorial did live in a biogenic paradise, surrounded by biogenic forces: the Garden of Eden, full of grass. Now what happened—that which theologians call the original sin: man started to think for himself and developed his own cerebral functions, and he believed that he now knew much more than the eternal biogenic powers that surrounded him—the tree of knowledge of good and evil, as it is mentioned in the Genesis. From that moment, man did not simply and instinctively absorb the eternal biogenic energies around him, but started another function—a limited, individual function—a cerebrospinal action, through which he was trying to live in a different way, figuring things out in his own limited way. So he lost contact with the Garden of Eden, and went on his own way.

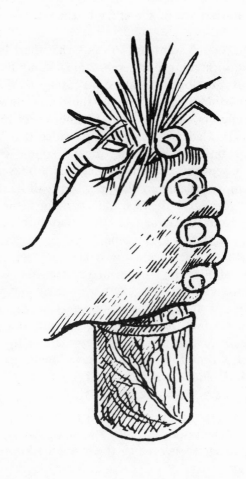

"... the Tree of Life becomes a tangible reality,
a biogenic battery between our hands. . ."

The Genesis originates also from previous ancient scriptures, and as such there are some symbolic, allegoric changes added to it— but this is the quintessence of the story in the Genesis, to answer your question.

Now once again, I don't want you to think that there is anything essentially wrong with using our cerebrospinal functions. Really, we *homo sapiens* developed tremendous brain power—we can figure out a lot of things—how to go to the moon, how to make complicated computers, and all that is fine—everything which augments human knowledge is wonderful, provided that meanwhile we don't become unfaithful to our original source of life and vitality and don't separate ourselves from this primeval biogenic force of the cosmos, because in that case we will be destroyed. We figured out wonderful rockets and thermonuclear machines that didn't come from the Garden of Eden, that didn't come from the leaves of grass, that didn't come from the Ocean of Life surrounding us—these machines came from our own minds—not our minds in harmony with the primeval biogenic forces, but from our own erroneous creation, our own erroneous ways—and this is the tragedy of man which may lead us to destruction, unless we return to the original, primeval force and infinite wisdom of life, which more than anything else is represented by the grass.

Well, I think it is time for you all to have lunch, because soon those beautiful dark clouds will gather and it will rain. I am always impressed by the rainy season here in Costa Rica because it keeps very regular hours—the mornings are full of sunshine and you can count on it that it will not rain until two in the afternoon. It is really a biogenic paradise here with this combination of sun and rain during the day.

*Professor, did you also arrange this climb up the hill every day for us?*

Well, you know, that hill is the department of Patanjali, the founder of the sixteen Yogas—he came about three hundred years after Buddha, in ancient India. He gave the right technique of breathing, and he criticized the different artificial breathing systems of India at that age—what he advocated as the most perfect breathing is when you walk on an incline of twenty-two or twenty-three degrees—this is exactly the incline of our hill here. Now you become automatically the followers of Patanjali when you walk up this twenty-two degree slope, that 250 yards from the highway to here. And when you do this, you have the perfect breathing system. All the wonderful things in the air here will supply plenty of oxygen to your brain, and you will be more receptive to ancient wisdom.

*. . . and so ended the fourth day. . .*

## The Fifth Day: July 30, 1979

We were talking in previous days about the "Greatness in the Smallness." I think it is time to say a few words about the fallacy of the "Greatness in the Bigness."

Very often we hear mentioned wonderful things such as the fact that 5% of the population of the U.S.A., the farmers, are supplying an abundance of food for the other 95% of the population. Well, it is true that the farmers make up only 5% of the population, but I want you to realize that after that 5% produced the foods, then about 35% more are busy to manufacture the tremendous amount of chemical fertilizers which the farmers are using, an enormous amount of insecticides and similar chemicals, then the great number of people busy in transporting those fruits and vegetables from one end of the country to the other, then the great number of storage places being maintained for those foods, then the processing factories, and the tremendous industry busy manufacturing artificial colorings, artificial flavorings, humectants, anti-oxidants, preservatives, and a hundred other things to put in foods, and then the canning factories, the packaging factories, all the industries which are busy with packaging materials, cellophane, plastic, paper boxes, and everything else, and then again transporting them to central storage places, and again transporting them to distributors and great supermarkets, and so on, until finally it arrives to the consumer, completely chemicalized, factorized, homogenized, deodorized, hypnotized—and there is born a product a hundred times inferior to whatever it was that was growing originally on the farmer's land. Now all these intermediary procedures in all their complexities are keeping busy 35 million Americans—in other words, 5 million are producing the foods on farms, and 35 million are taking care of all these intermediary nonsenses and superfluities so that the consumer can finally get that food. But it doesn't end there. The consumer still has to drive to the supermarket and drive home, so that in addition to the multiple transportation of all these things necessary to the food industry, preceded by all the incredibly complex things done to the foods themselves, the consumer also contributes by polluting the air with noxious fumes and wasting energy. It reminds me of a Latin proverb—you know, I always like my old friends, the Romans—they had a proverb for every situation in life—*pariuntur montes nascitur ridiculus mus*—the mountains labor and bring forth a ridiculous mouse. What a tremendous waste of human energy, of time, of labor, in order to deteriorate the products of 5% of the population, finally getting it to the table of the consumer, so the other 95% can ruin their health by eating it.

Now when you have a self-subsistent creative homestead, when the fruit or vegetable you planted is ripe, you simply go to your garden, pick it, and eat it. You are the producer, and you are the consumer, and you have eliminated 35% of all this superfluous labor, time, energy, pollution of the environment, and waste of our natural resources. This is the fallacy of the bigness and the value of the smallness. Nevertheless, television, radio, magazines, newspapers, all proclaim this fallacy, telling us what a wonderful country we have—5% of the people are supplying abundantly food for 100%—250 million people! If you analyze it, you will find that *contra facta nihil valent argumenta*—here come my old Roman friends—against the facts, arguments have no value.

It is not enough all this mess we are making in the United States. This mammoth food industry has to obtain almost all the raw materials from agricultural countries of the third world, like Africa, Asia, and Latin America—what is happening there? Well, it is a very sad story, one which begins in the United States. There, we see that finally, after decades of use, maybe 1% of these dangerous chemicals used in food production are discovered to be carcinogenous—cancer-causing. And this happens only after so much noise from the actions of honest physicians, groups of ecologists, consumer-advocates, and so on, that finally there is no way for the government to ignore the issue any longer. So they outlaw perhaps one or two per cent of these harmful substances. Then the gigantic fertilizer and insecticide industry has on its hands millions of pounds of these substances which cost a lot of money to manufacture, and suddenly it is against the law to sell them in the United States. It is then they have the brilliant idea to dump all these outlawed chemicals in Latin America, Asia, and Africa, poisoning millions of poor people who are being brain-washed and told that in order to efficiently produce foods, they must have chemical fertilizers, insecticides, herbicides, etc. First they brainwash the American public with the fallacy of the 5%, and now they are brainwashing poor farmers in Asia, Africa and Latin America, inundating them with agents, propaganda, etc., convincing them that their methods which worked for hundreds of years are no longer any good, and that to get the maximum results they must use from now on this fertilizer or that insecticide.

Of course, every cause has an effect, and karma catches up with them. A great deal of raw material—bananas, cocoa, coffee, beef, fruits and vegetables of all kinds—are being processed in countries of the third world. And naturally, in the production of these foods they are using very extensively all those chemicals which have been prohibited in the U.S. So all these outlawed harmful chemicals which were exported so ingeniously from the U.S. are returning to the U.S. as an integral part of the coffee, the beef, the fruits and vegetables. As one of my old Roman friends, Juvenalis, said, *dificile est satyram non scribere*—it is very difficult not to write satire.

But it is more than satire, it is tragedy. By using these concentrated commercial fertilizers they are using up their valuable top soil—the friendly bacteria in the soil, which are necessary to make possible the absorption of minerals by the roots of the plants—are being destroyed. It doesn't matter how much fertilizer you put into the soil, if you don't have this beneficial bacterial life, the plants will be impaired, the roots will not be able to absorb the necessary nutrients. So not only are we deteriorating the health of all these people by exporting illegal chemicals, illegal fertilizers, illegal insecticides, herbicides, etc., but we are also depleting the valuable top soil of Asia, Africa, and Latin America by destroying the bacterial life with these concentrated commercial fertilizers. We are not only foolish in our own country, in the U.S., but we are exporting our foolishness to the rest of the world.

We may be the leaders of the free world, but we have certainly lost the down-to-earth common sense of perhaps the greatest American thinker, Thomas Jefferson, who not only wrote the Declaration of Independence, but also introduced most of the best-known fruits in the U.S. from Europe—apples, grapes, and many others which most people don't know about—perfecting them at Monticello, from where they propagated all over the country. He wrote against bigness, against concentration, against the "pestilential cities." He prophetically warned that while our emphasis is on the small landowner, the small farmer, we would have real economic democracy and freedom. But if we would shift our center to the cities, to industry, we would eventually lose our greatness which was based on smallness, on self-sufficiency.

Of course, it turns out he was right, but it is small consolation.

Well, I wanted to explain these things to you in case you sometimes are tempted to let yourselves be brainwashed with these fallacies of propaganda. Don't believe it: there is no greatness in the bigness, because after something starts to grow too big, it brings diminishing results. It also brings ponderous bureaucracy, overlarge, inefficient government, overpowerful big business, equally big labor unions, corruption, etc., and this is why the world economy is getting sicker and sicker and why we have all these troubles around the world. Inflation, higher and higher taxes, recurring recessions, depressions, social unrest; economic unrest, all these are diseases of bigness.

So you see that my philosophy of the Greatness in the Smallness is based on facts. I first advocated this philosophy in the late twenties, when my book *La Vie Biogenique* appeared. Later on in England in the early thirties I wrote *Cottage Economy* on the same subject, and in the forties, in the U.S., *Father, Give Us Another Chance*. And just recently, because I am a very tenacious person, you know, I wrote *The Greatness in the Smallness*.

I want you to realize that there is no other way but to use common sense. Socrates, who was the greatest of the Greek philosophers, said that when we follow the path of reason, everything will turn out wonderful in our lives. In the moment we deviate from the path of reason, there will be everywhere a mess and catastrophe. So I just want to appeal to Socratic reason, which we shall try to apply in all things we are doing. Now, what is this Socratic reason applied to our concept of living? Let's get down to basic essentials. Let us say you finally get an acre or half an acre of land. In a northern country where there is half a year of cold, you need a full acre. In a good climate, say a southern climate like we have here, where there is no frost and you can grow things all year round, you need only half an acre. So let us suppose you have that acre or half an acre of land. Following the path of reason, you cannot just plant a few seeds there and expect to have good vegetables. You have to start in the beginning, like Zarathustra started in the beginning with the Creation. And you have to continue the work of the Creator on that little piece of land by first creating your top soil.

I mentioned before that our whole civilization is built on that few inches of top soil, perhaps the top twenty inches. If that is lost, we will starve, in spite of all our tremendous technological achievements. Well, usually when you acquire a piece of land, there it is, probably full of weeds, and due to the action of wind and sun, probably very dry. When moisture disappears through evaporation, all the wonderful friendly bacteria in the soil which make possible the creation of life, also are destroyed. Therefore, you have a gold mine in the moon, because although your land is there, you cannot grow things. Here you have another erroneous belief: somebody thinks, well, I will buy a little piece of property and plant seeds and live off the land. No, first you have to continue the work of the Creator. In Genesis, the first thing described is the formless void of the universe before the Creation—in the original language, *tohu va bohu.* So you have to start the Creation. You cannot say, "Let there be vegetables!" Nothing will appear. First you have to create top soil.

You have tremendous allies when you begin this task: the sixteen forces. You don't even need all the sixteen, just a few. And these few, working with the laws of nature, can help you make compost, which is the first step toward creating a fine top soil. The Latin proverb says, *omnia orta exeunt.* Everything which appears, disappears. Everything which appears, decays. That weed which appears on your land and which you think is an enemy, a useless thing—not at all. You cut it up, it decays, and you have made a first step toward creating a top soil. Every time you eat fruits and vegetables you save the peelings, you save all the waste products of the healthy things you eat. You make a layer of your kitchen waste products, a layer of soil, a layer of cut-up weeds, and again a layer of garbage, soil, and weeds, and in a few months, without any complicated machinery, you will have good black humus.

You can simply pile these things up and eventually they will decay and make a fine humus, but if you have a little more energy, you can do some additional things to increase the fertility of your compost and make sure it decays, and does not rot. First, have a little roof—any old thing, a simple roof so that the rain will not wash out the water-soluble minerals from your compost and carry them away. Then, you can avoid the work of turning the compost—because it must be turned two or three times to allow air to reach every part of it—by making a compost box constructed completely of laths, even on the floor, so every side of the box

The compost boxes are made only of laths, so the compost inside is exposed to the air on all sides. The finished topsoil is removed through the door on the bottom. *(below)* Some of the garden beds in their first stages of construction.

is exposed to the air. When air comes in from everywhere, it stimulates the bacterial action, and the different temperatures in the box—below is cool, above is hot—create a kind of intensive metabolism, and you don't have to turn the compost heap by hand. Although it is good exercise, it is superfluous work, which we can avoid by using Socratic reason.

It takes time. For perhaps the first six months, instead of growing vegetables, you have to grow top soil. I say "grow" top soil, yes, because you have millions of friends: those friendly bacteria, who will help accelerate the decaying of your vegetable substances, weeds, garbage, and all these things, and contribute to the creation of your top soil. Only after you have spread the new top soil over your land are you ready to begin to grow vegetables, or plant trees. Because *ex nihilo nihil,* says the Latin proverb—nothing comes from nothing. You cannot expect to have vegetables and fruits rich in minerals from a soil which does not have minerals—you can take out from the soil only as much as you put into it.

I mentioned that the compost should decay, not rot. If you cut weeds and leave them out in the rain, they will become acid and start to rot. But if you always put the weeds in your compost heap under a roof, they will decompose. You should from time to time sprinkle a little water on it to keep it moist—that small amount will not make it rot, just decay. This process of decomposition will create a very high temperature on the inside—normal fermentation caused by millions of bacteria—the temperature will go up to a hundred twenty degrees. And in the end you will have a treasure—a beautiful black humus that will grow wonderful vegetables, fruits, everything you need.

You don't need to use one ounce of commercial fertilizer. Concentrated commercial fertilizers chase away and destroy the earthworms, and earthworms, beside of those millions of friendly bacteria, are your best friends. Earthworms are living on decayed vegetable material—not on material which has already decayed, or material not yet decayed—but material in the process of decaying. According to Darwin, in his wonderful book about earthworms, our top soil really was created by the earthworms, because their castings are the most perfect plant food. And when you create top soil naturally through composting, you will attract the earthworms—you will have hundreds of them.

You have many of the sixteen forces at your disposal in your microcosmos, in your garden. You have the sun, which will give chlorophyll to your plants; you have water, to give them moisture; you have air, which also supplies valuable ingredients such as nitrogen, etc.; and you have the earth itself. There are immediately four of the sixteen forces at your disposal, very anxious to work with you, and believe me, they are superior to the commercial fertilizers.

So finally you have good top soil. Then you are ready to plant. Not before. And you will have wonderful results. Then, as soon as your plants are growing up a few inches, you can practice soil conservation and conservation of water. If you just let your vegetables grow, all kinds of weeds will appear and will use up those valuable minerals of the top soil which you created, because weeds have more vitality than the plants you will grow. Remember the wonderful statement of Zarathustra—the *Zend Avesta* is really a gold mine, not on the moon, but here—he said that the real children of nature are the wild plants—the domesticated plants we grow are only nature's stepchildren. This is why you will see weeds all over, because they grow faster and with greater vitality than your vegetables. Nevertheless, in that small area where you have planted your vegetables, it is not convenient. You want to eat parsley, dill, onions, garlic, tomatoes, cucumber, all kinds of greens, etc.—you don't want to eat weeds. Not because it is not possible, but I don't think it will be very pleasurable. Therefore, you have to prevent weed formation.

Well, one ordinary way is just to go and cut the weeds. But that takes a lot of time, energy, and work. Or you can apply the principle of prevention—you can use mulch. Mulch can be dry hay, dry leaves, or in the worst case you can cut up your daily newspaper—at least put it to some good use. And then you spread the mulch between the vegetables, and that will prevent the weeds from growing. And you don't have to cut it or bend down or work hard. Some people use plastic sheeting, and it will prevent the growth of weeds. There is only one thing—if you use for mulching dry leaves or dry hay instead of plastic, in that case they will decay in time and enrich your soil; but the plastic will not decay nor enrich your soil. Therefore, it is better to use substances which decay, which are biodegradable, instead of substances which will not contribute to the richness of your top soil. So, after making compost in the simplest, most efficient and intelligent way, which needs very little work, the next important thing to learn is how to mulch your garden. Not only will mulching prevent the growth of weeds, conserving the minerals of that valuable top soil for your favorite vegetables, but it also achieves something else: it conserves the moisture of your soil, because it breaks the capillary action. When there is wind and sun—and they are everywhere—moisture is practically drained from the ground, a kind of invisible evaporation which endangers the life of those friendly, helpful microorganisms, for whom moisture is a basic precondition of life. You must always be considerate of these millions of friends who are working for you incessantly in the top soil, making it possible for the roots of the vegetables to absorb the nutrients. When by mulching you prevent the capillary action and the loss of moisture through the air, you can reduce your irrigation work by ninety per cent. You will have to use only ten per cent of the water, which is an excellent thing considering the general water shortage in the world. Remember that pumping water uses energy also, as well as machinery, as well as physical labor and time. Therefore, mulching aids in the conservation of the soil, the conservation of the nutrients of the soil, and the conservation of the moisture of the soil.

Then of course, in your great enthusiasm you will not go and plant a whole little envelope of seeds all at once. If you do that, in two or three months you will have a tremendous harvest of radishes or carrots or onions, and you will not be able to eat one-tenth of it—the rest will be useless. Here again is the principle of the greatness in the smallness. You should plant now a small line of onions, then in two weeks you plant another small line of onions, and do the same with the other vegetables. They will come up gradually and will supply you continuously with fine, fresh organic vegetables, and in the measure as you are eating them, they will continue to grow. You will not need refrigeration to store them, you will not have to freeze them, you don't have to do any of those superfluous activities. When you need something, you just take it from the garden and eat it. According to the Plinius manuscript, this was the practice of the ancient Essenes. There is no better place to preserve all the vitamins, enzymes, plant hormones, everything that is good in a vegetable, than in the ground where it is growing. In its natural element, it has all the biogenic energy of a fresh, growing food. And then you eat it. And you will have ten to fifty times more nutritive value than something which was harvested in the other end of the country, then transported, stored, refrigerated, and who knows what else before finally reaching you. And not only do these delays cause deterioration of vitamins, enzymes, etc., but in the process the vegetable or fruit is increasing tremendously in price. A bunch of onions which may cost three cents to produce may reach forty cents by the time the poor consumer is buying it. The economic factor is also involved; even from the viewpoint of economics the smallness is more efficient than the bigness. But everyone is brainwashed and believes that the only way to get a vegetable is to go to the garage, start the car, drive a few miles to the supermarket, waste an hour of time, stand in line to pay for it—ten times more than it is worth—drive home, use more energy, pollute more air, then put it in the refrigerator. Even this is not the end, because that poor vegetable, already containing only a fraction of its original vitamins and enzymes, may well be subjected to further atrocities by being oversoaked, oversalted, and overcooked.

Well, I may mention a few other things. When you grow your vegetables, it may happen that you will have competitors: other creatures who also like your vegetables—all kinds of insects, for instance. Well, here again you are brainwashed—you think you have to go to the store and buy insecticide. It will probably kill your insects—but you will eventually eat that vegetable and it will do biological damage to yourself—and after a few generations of insects are killed, you also gradually will be debilitated, and all kinds of symptoms of all kinds of ailments will appear. Therefore, it is not a satisfactory solution. Analysis of the tissues of the average American shows an unacceptable amount of DDT in the fatty tissues, fifteen years after DDT was outlawed. Mother's milk also shows a high level of DDT, and that is only one of the insecticides—I could mention a few hundred more. So that is definitely not the ·solution. So how will an organic gardener defend his vegetables against insects? *(one of the participants mentions the use of frogs)* Yes, frogs are excellent friends in your garden; they eat a tremendous amount of insects. But you shall not depend entirely on frogs—you cannot go and talk to the frogs, telling them, please come to that plant and eat those insects around it— nor can you call a swallow from somewhere and say, now listen,

179

go to that tomato, I have some insects there! You need some quick action. All right. There is a wonderful vegetable called garlic. You liquefy garlic with water, spray it on the plant which is affected, and it will take care of your insects beautifully. Nothing but a concentrated garlic solution. There is a very interesting thing in garlic, called *oleum allii,* a kind of essential oil, which destroys harmful microorganisms. It also destroys harmful microorganisms in the human colon if you eat it regularly—I mean, not garlic alone, but adding a little each time to your vegetable salad. It does the same thing to your harmful microbes as to the insects which are attacking your vegetables. Then there is that strong chili—that very hot pepper which is grown in Mexico in abundance, and which has a counterpart in every country—you practically cannot put it on your tongue because it will burn it. A very small amount is stimulating to the gastric juices and is healthy if you put it in your vegetable salad or your cereals. But when you liquefy it with water and spray it on your plants, it will do the same thing to the insects as garlic. To a great extent, even onion will do. And if you liquefy together garlic, chili, and onion, you really have the atomic bomb! You spray that over your plants and you will have no problem.

Then there are your fruit trees. All kinds of little things like to crawl up your trees and into your fruits. Well, you apply the principle of prevention. You can use different things. Sometimes in a mild case you can make a simple lime solution. You get ordinary lime and liquefy it—put a tablespoon of lime to three quarts of water, and just paint the lower trunk of the tree and around the tree. Or you can use soap, just ordinary soap, and put it on the tree. Or you can use number ten motor oil—no ants, no insects will go near it. Or you can combine all the three things.

I have a winter home at Lake Chapala in Mexico where I set up an organic garden. And as I am traveling all the time and am very little there, I taught my gardener how to take care of the fruit trees. My method was this: first we made a ring around the tree on the ground with a little cement, or with bricks. Then at a distance of about six or eight inches we made another ring. And between the two rings we made a little cement floor and put there water, and on top of the water a little kerosene. We never had any trobule with ants or insects because they cannot go through the water. You can eliminate about seventy per cent of the invaders in that way. (I may have been inspired by the moats around the castles of my Transylvanian ancestors!)

You can also put up some bird houses, with a little water, and attract the birds. They will catch those flying insects before they reach your fruit trees. An average swallow eats about five hundred insects a day. And I already mentioned frogs.

So there are very simple ways to save ninety per cent of your fruit and vegetable crop. And don't worry about the other ten per cent. Remember that insects were here millions of years before man on this planet, and they have some priority rights. So I think we can let them have ten per cent. After all, we are the intruders— we came much later. In some kind of a cosmic courthouse or tribunal, we will lose if we take our case against the insects. If you save ninety per cent of your crop, you will have much more than you will ever need.

Sometimes the simplest things are the best—again the greatness in the smallness. When you have a small vegetable garden and you walk through it every morning for ten minutes, when you see an insect, or a worm, you just pick it up and take it away. You can have a little can and put them in it, and then when you have a little harvest of a few dozen worms or insects, just take them away and burn them. They will never come back. In fact, you will have some ashes which will be very useful for your compost. You can do this in a small garden, but imagine a big food-producing company with fifty thousand acres of some plant—they cannot do it. But a small farmer, who has just a few square yards of vegetables for himself and his family, can easily do it. Or you may have aphids on your leaves. Just get your watering can—you don't need a complicated irrigating system—and just zoom some shower water there and it will wash the aphids away. Once they are washed off the plant they do not come back. And there are also beneficial bugs who eat other insects, such as the lady bug, and the praying mantis. You can throw some in your garden and they will eat a tremendous amount of harmful insects. There are such simple ways to control insect pests, without insecticides. It is very interesting, the wisdom of nature. Life forces exist in the insects, too. They also have bioactive power. After a few generations of being attacked with insecticides, they develop immunity against the substances, and soon come species which have immunity and propagate gaily all over your garden, in spite of the insecticide.

This is the intelligence of the genes of the insects. Insects also have DNA and insects also have this biogenic heredity for millions of years, and in order to survive, the inborn intelligence of the genes enters into action, changes the biochemistry of the insects, and they survive. So after a few generations, the insecticides will have no value. But meanwhile, it has done a lot of biological damage to your tissues and your basic biological functions. Therefore, here comes again my old friend Socrates: if we walk on the path of reason, everything will turn out well in our lives. But—in this case—if you are brainwashed by the insecticide people, everything will be a disaster.

*(Norma asks Professor to tell how he once used an ingenious method to get rid of tomato worms.)*

Oh yes, that was very interesting. It happened at Rancho La Puerta,* where we had organic gardens and hundreds of acres of vineyards, for the practice of a diet of grapes for several weeks for the purpose of general cleansing, and improving the chemistry of the blood—raising the hemoglobin and erythrocytes, and so on, as these grapes were grown on very fertile soil by organic methods. We also had about a hundred acres of wheat, and extensive organic gardens—we were self-subsistent—as we had to feed about two hundred people, guests and also workers. So one day there came my gardener and told me that we were in trouble—our two acres of tomatoes were infested with those large green worms, and he wanted to know what to do. He knew my philosophy about just picking up worms and burning them, but it would take a year to collect so many. I told him to make a lime solution and spray the whole crop with it. So he went off to carry out my instructions and two days later came back with a very long face, telling me the tomatoes were infested much worse than before. Well, those worms probably had a calcium deficiency and with the lime solution were growing much faster and more vigorously! So now I really started to meditate seriously. I thought, after all, we cannot let these inferior creatures which, according to Zarathustra, belong to the kingdom of darkness, defeat *homo sapiens!* We must use some intelligence, some Socratic reason. Now one of the products of my vineyards was a grape vinegar—I had several barrels. I told him, listen, take one barrel of this grape vinegar and spray

*See *Search for the Ageless, Volume Two: The Great Experiment.*

Professor gives a tour through the banana and plantain orchard of the I.B.S. Center. *(below)* At Professor's winter home on Lake Chapala, Mexico, cement rings around the fruit trees effectively keep ants and other insects away.

all the tomatoes with it. So he went off again, and the next day he returned triumphantly, saying, "Professor, we don't have one living worm! They are all gone!" Of course, we ate pickled tomatoes for about a week, but at least we saved them. So I just give you this example to realize that it is possible to use natural, simple methods to get rid of your worms and aphids.

*Professor, I heard Earl Butz say that if we were to use only organic methods, who would decide which fifty million people in the United States would starve? What do you think of his statement?*

Now here again is brainwashing. It is very simple: mankind for thousands of years grew foods and survived without the chemical industry, which has existed for only the last fifty years. Now if Mr. Butz were right, in that case mankind would have been dead thousands of years ago. The statement really doesn't stand up to Socratic reason at all. Thousands of organic gardeners who grow their own vegetables and fruits could testify that he didn't read *The Dialectical Method of Thinking,* and he surely didn't study Socrates.

Now I just wanted to give you a glimpse in a nutshell about the "smallness of the bigness" of the food industry, the chemical fertilizer industry, and the insecticide industry in general, and the greatness in the smallness of the small farmer who is using natural methods. And I want you to meditate about it and never let yourselves be brainwashed by the food industry or the chemical industry. Just go back to our old friend Socrates.

*What about marigolds?*

Marigolds are fine plants. Insects don't like marigolds and will not go where marigolds are growing. You shall not imagine that marigolds go around chasing insects in your garden, but somewhat the flower repels insects—there is something in it the insects don't like. Here I forgot to mention a very important thing. Of all these natural pest controls which the modern contemporary organic gardener tried and experimented with, there is one which is the simplest of all—one which has been forgotten for thousands of years because it is one of Zarathustra's methods in the *Zend Avesta*—I repeat, he was a genius of simplicity. It is the simplest possible thing. Do you know what advice Zarathustra gives to the gardener who has trouble with insects? He said to walk around in the neighborhood and observe which weeds are growing most profusely, with the greatest vigor, and which are not bothered with insects. Then chop up those weeds, mash them and mix them with water, and throw the mixture over your plants and all around your plants. He pointed out that weeds in the vicinity of your garden which are not attacked by insects must contain something the insects don't like. It is evident, otherwise they would attack them. And it really does work. From time to time, about every two months, gather some of these vigorous weeds—the real children of nature, according to Zarathustra—chop them up, triturate them with water, and sprinkle them over and around your plants. That will take care of your insects. I tried this out in many places, and it works. Nobody yet discovered it in the twentieth century, not even organic gardeners. And it cannot be more simple.

I also want you to think about this. If your biochemistry and metabolism is healthy, if you have no deficiencies, if your bloodstream is biochemically well-balanced and vital, then some bacteria which may get into your system will have a very difficult time and will be liquidated by your phagocytes and your lines of defense and you will not get ill. It is exactly the same thing in the vegetable kingdom.

*Why were the Indians of long ago, who were so healthy and vital, decimated by white man's diseases?*

They were not decimated by disease, but by alcohol. It did tremendous damage—to the liver, to the hormonal function, to the bloodstream, to many basic biological functions. Then also there were other factors which were not insignificant—for example, the firearms of the settlers who righteously liquidated those Indians. I remember that wonderful statement of Voltaire, who said, "The missionaries arrived to the New World. They had the Bible. The Indians had the land. And a few years later, the Indians had the Bible, and the missionaries had the land!" The Indians were dislocated completely. Their way of hunting had been to kill one animal or two or three, only what was necessary for them to eat in order to survive. They didn't go to massacre hundreds and thousands of animals to sell their fur as the newly-arrived colonists did. Therefore, little by little the Indians were pushed back farther, and little by little they had less and less food. This is what liquidated them, not diseases—outside of the alcohol, and syphilis. They had no natural defense against the *Spirochaeta pallida.* They died by the thousands from syphilis and alcohol, both of which we imported from Europe. They were weakened and had less and less food because their natural environment was destroyed. But though the white man gave the Indian alcohol, the Indian got his revenge by teaching the white man how to smoke. The Indians were smoking only on rare festive occasions, but the white man adopted smoking as a habit, with very unfortunate results, as we all know.

*It seems to me that correspondence can be handled from anywhere in the world. Why did you choose Costa Rica for your (I.B.S. International Correspondence) Center?*

Letters arrive all the time asking this same question—from Australia, from India, from all over. There are several reasons. I have already mentioned the obvious ones—that our philosophy is based on the *Essene Gospel of Peace,* and Costa Rica is the only country which has no army or munitions factory—where peace is a way of life. But there is also something else.

During my archeological fieldwork in the south of Mexico for many years, I developed a keen interest in Latin America, and the person with whom I collaborated in my archeological work was the founder of the University of Guadalajara, Licenciado José Guadalupe Zuno. He was from an old family—his ancestors came to Mexico with the *conquistadores*—and he graduated from the Sorbonne, which was also my alma mater. He was an excellent artist, he was Governor of the state of Jalisco, he wrote several books on different subjects, and he was the greatest proponent in Mexico of my ideas about archeosophy—the combination of archeology and philosophy. You may know him better for one irrelevant thing—he was the father-in-law of the previous President of Mexico, Echeverría.

As the founder of a great university, he was always interested in education, and after becoming acquainted with my archeological work, he set before me a very interesting proposition. Knowing that I had studied the Indian population in remote mountain areas, he asked me if I would recommend measures to improve their health, their standard of living, and especially the nutrition and hygiene of their children. He told me that if I would write a book about my findings, he would see that my recommendations be put into practice, and he had the means, as his daughter was First Lady of Mexico. So I did as he asked, and the result was my book, *Los Pastorcitos*—The Little Shepherds—with the sub-title, *La Salvación del Niño Campesino*—the salvation of the peasant children. The sub-title was not an exaggeration, because I really found tremendous deficiencies among them. But it was not enough to write a book, and not enough even to have the power of the Presidency to put my suggestions into practice. Because there is a great handicap in Mexico: a high rate of illiteracy. Therefore, a program of education would have taken many, many years—maybe not finished in our generation. So I had an idea. The Mexican government has all over the country little stores where foods and things for general consumption are sold at cost price to poor people—they are called Conasupo. So my suggestion was this: a great number of Indians in the most remote areas were living practically on corn exclusively. The tortillas made from pure corn are good food, but corn has no complete protein—it doesn't have all the essential amino acids. Therefore, I suggested that the government acquire the machinery necessary to mix together soy powder with the corn *masa*—that dough-like substance which is the corn after it has been soaked and ground—a proportion of ten per cent soya and ninety per cent corn. That ten per cent soya has the essential amino acids which are missing in the corn, and the combination of the essential amino acids of the corn and soya together makes the whole mixture a complete protein. And there is no difference in taste, so they would not have to tell to the stores, or to these people who buy the *masa* from them that anything is different in it—it would be very difficult to explain amino acids to simple people who are illiterate—just simply sell it to them as the usual corn *masa*. And in that case they would automatically have no more protein deficiency. And this is exactly what happened. Their ordinary tortillas became extraordinary, as the soya transformed them into a high-protein food which I baptized the "tortilla rica." To solve the problem of hygiene, I also recommended the use of the *temazcalli*, a custom practiced by the Aztecs and Toltecs in ancient Mexico—a little adobe room where a quart of water is boiling on a small wood fire so one can take a vapor bath. These poor people have no plumbing, and they cannot afford to establish bathrooms, but the *temazcalli* is something anybody can build—it doesn't spoil the ecology, it needs only a quart of water and a handful of wood. And I made several other recommendations in that book, which were carried out very successfully.

Because of this experience in Mexico, I was really impressed to discover that in Costa Rica there is practically no illiteracy—more than ninety per cent of the population reads and writes. I thought it would be good fertile soil to try and introduce my ideas about organic gardening and similar subjects. Here there would be no problem with communication, because not only is there almost universal literacy, but reading seems to be everyone's favorite sport. So I wrote a little book about the precolumbian ball game, and how to reconstruct it in schools and colleges, so the young people of today could not only enjoy a healthy sport, but also become aware of the spiritual and philosophical heritage of their country. I gave the book to the President of Costa Rica, who transferred it to the Ministry of Culture, and they are printing it now. Also, I am preparing a book to try to remedy a certain condition, which unfortunately is prevalent in many parts of the world, not only Latin America. You see, the peasants of Latin America don't know about organic gardening for two reasons: one, organic gardening does not consist only of growing foods without chemicals. Unless they know also the technique of how to make compost, how to mulch, how to control insects by natural methods, and a few other things, they are not able to use their land efficiently. Two, they are brainwashed by the chemical fertilizer and insecticide industries, and of course, these poor people who have just two or three acres have no money to buy chemical fertilizers and insecticides, nor the complicated irrigation systems they want to sell. Therefore, they get discouraged. They don't realize that on one acre of fertile land with the right methods they can grow more foods than they could grow on ten acres otherwise. So then they get discouraged at not being able to buy those fertilizers, insecticides and irrigation systems, and they sell or abandon their land and go to the cities to try to find work. Then the cities are growing, unemployment is growing, the crime rate is growing, the ecology of the city is deteriorating, the expanding population needs more plumbing, more hospitals, more schools, more of everything, and all these strains create serious problems in a poor country. So this is my idea. The peasants are working six days a week. They have one free day, Sunday. If it would be possible to organize a kind of seminar on ten consecutive Sunday afternoons, each about two hours, to teach them the right natural methods of gardening, and to teach them they can grow on one square yard of land, made more fertile by organic methods, more food than in twenty yards of the neglected way, then they will not need chemical fertilizers, nor insecticides, nor any complicated system of irrigation by using natural conservation of the moisture in the soil.

One afternoon, the Secretary* of the President of Costa Rica brought over the head of the land and development section of the Department of Agriculture to see me, and we had a long conversation. Both of them were very receptive to my ideas—he said I should write a book about them, and he would see to it that it was printed by the government and they would mobilize all good people who were willing to try them out. This solves not only problems of food production, but social and economic problems as well—to prevent the proletarization of the peasants—to prevent them from leaving rural areas and going to the cities trying to find work. And of course, it improves the balance of payments of the country, as it doesn't make much sense to grow certain crops, to sell them to the U.S. and other countries, and then pay high prices in the market for foods of inferior quality which they could grow themselves of superior quality for their families at much less expense. So this is one of the projects of the I.B.S. which we want to work on, as Costa Rica is an excellent country to make an example for the whole Latin America, because illiteracy is at a minimum here—over ninety per cent of the population reads and writes—and it has great social and economic importance. So this is another reason why I am staying here longer these days than I am in the U.S. or Canada.

We have other interesting projects. We are establishing little libraries of Essene-Biogenic literature in prisons in the U.S., for example, and also in some other countries. The purpose of prisons really should be the rehabilitation and re-education of the prisoners, but not much is happening along those lines, because after serving their terms, they usually leave the same embittered person without hope for the future as when they entered. The foods in prisons are usually not very biogenic—the prisoner's health definitely does not improve there. But we in the I.B.S. feel we can do something to help, and so we send literature there, Essene-Biogenic literature—several wardens are very cooperative, as they

*Sr. Juan Vicente Lorenzo, who is translating many of Professor's books into Spanish, including *The Living Buddha* and *The Essene Jesus*.

have plenty of trouble with riots and they are delighted when these people are reading the *Essene Gospel of Peace.* And there is another thing. A great many of the prisoners have made little boxes and grow greens in their cells to supplement their deficient prison foods. These positive things cannot help but change their outlook, and they are able to see the possibility of starting a new life after they leave the prison. Many of them are studying nutrition and the right way of living—many of them want to practice and even teach these things in the future. Another project is sending books behind the Iron Curtain, to countries which are under dictatorship, and to Third World countries where they cannot afford to buy them. In fact, we are using all the revenue of the books we are distributing, to do this.

You probably noticed one thing: our books are never advertised anywhere. Books usually which are published by publishers and sold in bookstores are advertised—they put advertising in trade magazines and other kinds of journals—we never do that. It goes only by word of mouth, person to person—somebody reads the *Essene Gospel of Peace,* tries it, somebody reads the simple book on living foods,* tries it, somebody reads our guidebook, *The Essene Way-Biogenic Living,* tries it, their friends read them, they are trying it. In this way we don't have a tremendous number of people who through curiosity will go and buy a book in the bookstore—but they have an incentive to buy it because they feel it satisfies a certain need. Without any advertising, the *Essene Gospel of Peace* now reached eight hundred thousand copies published only in the U.S., without speaking of all those published in England, in Germany, even in Japan, and in about two dozen other countries. It has been translated into twenty-three languages, as have the other books. So we are trying to be as active a point in the universe as possible.

The function of our International Biogenic Society has many aspects. First, it coordinates the publishing programs of our different publishers in various countries. Second, it also establishes a bridge between ourselves and the readers of all these publications. The publishers carry on extremely useful work, publishing the books and distributing them to bookstores. But the readers cannot ask questions from the publishers, because in most cases they concentrate only on the printing and distribution of books and

*The full title is *The Book of Living Foods.*

190

are not acquainted with the scientific, archeological, historical and spiritual aspects of these teachings. Therefore, when letters arrive, as they do all the time, from the readers to the publishers, they are transferring all these letters to our International Center of Correspondence. Then the International Biogenic Society also sees to it that in many countries the members of the I.B.S. get discounts from these publishers, which is very useful for the members.

Another aspect of our functions is the propagation of our literature, and especially the *Essene Gospel of Peace,* in the countries behind the Iron Curtain. You see, pacifism is a wonderful thing, but as Anatole France said, pacifism unfortunately appeals only to pacifists—this is its weak point. Therefore, it will be useless for us to propagate the philosophy of pacifism in the western countries only. Unless we can penetrate behind the Iron Curtain we are doing only half the job. It is really extremely interesting how we successfully smuggled hundreds of copies of the *Essene Gospel of Peace,* for instance, to Russia. There it is being translated into Russian, typed with old typewriters, then printed on a very primitive machine where the typed paper is put on some kind of jelly and copies are run off—they cannot make more than fifty copies at a time. Nevertheless, along with the underground literature, a great number of the *Essene Gospel of Peace* are being passed from hand to hand, often at considerable danger for those who are doing it.

We are also sending out a good number of our books as presents to countries of the Third World where people don't have the means to buy these books, a good number of countries in Asia, Africa, Latin America—we have a very extensive correspondence with these. We are continuously sowing seeds all over the world.

Then there is another aspect, not necessarily only to the Communist countries, but to a great many countries where people live under dictatorship—we also send there, to a great number of people who are extremely grateful for our literature. You have to realize that eighty-six per cent of mankind is living under dictatorship—only fourteen per cent of humanity live in free democratic countries. Of course, democracy is not perfect. Anything which is governed and managed by human beings is not perfect. Nevertheless, in spite of that, it is the best of all existing systems.

Now Costa Rica is one of these democracies. It is amazing,

surrounded by dictatorships, how from the beginning Costa Rica is a democratic country, with free press, free elections, and a very liberal constitution, with all observance of human rights. But in addition to all this, Costa Rica has one special great value: it is the only country in the world which has absolutely no army, and no manufacture of munitions. This means that pacifism here is not a philosophy, a theory, but a way of living. It is for this reason that two years ago we united our different correspondence offices from various countries and brought them here to Costa Rica, establishing here our international correspondence center. We have an extremely efficient post office in Cartago—since we are here they have to work ten times harder, because every day arrives a big pile of correspondence and this office is sending out at least fifty letters a day to different countries. Nevertheless, they don't complain—they take it as a matter of fact that this strange institution suddenly appeared and letters are coming in from the most exotic countries in the world. My ideal was always for an international correspondence office not to be somewhere on the twenty-seventh floor of a big air-conditioned building downtown in noise and polluted air, but somewhere outside where there is fresh air, sunshine, vegetation, trees, and peace. So this is why we are here.

Even the office of the I.B.S. represents in action my philosophy of the greatness in the smallness. There is the whole international correspondence center in two rooms: one is the study room with the reference books, and so on, and the other is a little office with two IBM selectric typewriters, with the different fonts so we can correspond in a dozen languages; and one IBM Composer—a kind of pocket battleship, a little printing plant in itself which creates with the greatest ease photo-ready copies, ready for the printer. This was necessary to have because very often in the past when one of my manuscripts arrived to the publisher, they would typeset it, send back the proofs, I would correct the proofs, and then in the process of making them over they would make new mistakes because of all those Latin and Greek words I am using in my archeological books. So it was an endless procedure—in this way they get photo-ready copy and they can go to press right away with it. Then we have a large copy machine which makes a copy of every letter we are sending out, and every manuscript we are sending to different publishers. If one should get lost in the mail, it is not lost forever. We also have a file of thousands of members all over the world. And all this in one room, which is very comfortable in spite of being the center of so much activity— this is truly the greatness in the smallness.

I always told to those who work for the I.B.S. never to grow too big—it is possible to perform all these tasks simply from a small place out in nature, away from the city, so they can give an example of the Essene Way of Biogenic Living, which is very difficult in a noisy, polluted city. I hope the I.B.S. always will represent the greatness in the smallness. Although I founded it with Romain Rolland in 1928, as the spiritual head of the movement, I don't interfere much with the practical management. But I still have enough influence, and I think it will be kept small.

*"... always remember, the greatness is in the smallness ... "*

Today I want to dedicate to your questions. The value of a Seminar is that you can ask questions, that you can have a dialogue with me. You can read my books at home—if all you want is what is in my books, you don't have to make a trip of hundreds of miles to get here. The main thing is that here you can talk with me. You see, there is no better way of imparting knowledge than through Socratic dialogue. Socrates had the right method: questions and answers. It is the most important thing: we cannot learn any other way as well as by questions and answers. Therefore, I want to give you all the opportunity to ask me questions. I prefer not to talk about things I am interested in, but things which you are interested in. Our whole culture is based on Socratic dialogue—between man and government, between man and God—always a dialogue which is going on—and the questions and answers create a synthesis which carry us nearer to a solution. Therefore, I prefer to answer all the questions you have. And remember, please, the most important thing in a Seminar is that you can ask questions. You cannot ask questions from a book—and unfortunately, I wrote over a hundred books. Now this is a violation of a law of Zarathustra when he said that if we learn how to read from the book of nature we will not need any other book—so if I ever meet good old Zarathustra, he may not be very happy with me. But I feel that in the twentieth century the only way to counteract the obnoxious effects of television and radio and newspapers and advertising, is through good books. I believe in books. Books are not noisy, like all these technological methods. Books can be read quietly, and a book also has the great advantage that when you want to stop reading, you just close it, and you can meditate. I believe in books. Nevertheless, a book has only as much as is printed on its pages, but when you talk with me, you can ask all those things which you didn't learn from the books, and maybe many other things you are interested in—therefore, don't hesitate to ask questions.

*Professor, what do you think of faith healing and revelations received by going into a trance?*

Well, it has a serious aspect and a humorous aspect. Let's talk first about the serious aspect. Now there is a very important point concerning the sixteen forces of Zarathustra. He wants us to be receptive to these forces around us—and he also wants us to acknowledge the reality of these forces. You remember the Ahuras of sun, water, air, food, and so on—he wants us to be conscious of the reality of these forces—and not only to be conscious of their reality, but to be receptive to them, to be able to receive them as realities. Zarathustra wants us to have faith in the reality of things which *are*. The sixteen forces of Zarathustra are evident, we can touch them, we can experience them. Now very often appears some kind of movement which is based on faith. Unfortunately, it is not based on the reality of certain forces, but on artificially created abstractions, not corresponding to the reality. Nevertheless, in spite of the fact that this belief usually is not based on reality, whenever an individual develops a strong faith, that act causes the organism to throw into the metabolism and the biochemistry a great number of enzymic and hormonal factors which have a healing effect. If you develop a strong faith in anything—even a box of matches—and you believe that box of matches has divine powers, the organism will throw into the biochemistry very powerful enzymic hormonal factors, and healing may take place. But of course, the superiority of the Zarathustrian sixteen forces, or the Essene Angels, is that *they correspond to the reality,* which is tremendously important. When faith corresponds to the reality, and that faith is powerful, it will result in healing.

But here is the most important thing I want you to realize: remember in the *Essene Gospel of Peace,* after Jesus the Essene healed the sick, he said to them: Go, and sin no more. Remember these words. Suppose someone is healed from a disease by a strong inner spiritual experience—faith—and because of that faith, certain powerful enzymic hormonal factors were thrown into the system and healing took place. But that is not enough. If that person who got healed will go back to eat the wrong foods, follow the wrong way of living, continue to smoke and drink, etc., he will continue to sin and then he will be sick again. This is why Jesus said, go, and sin no more. He explained about the Angel of Food,* the Angel of Water, the Angel of Air, all these things—how to live, and how to heal in a simple way the material plan—because he knew that very few people would have that receptivity to get healed by

*Because of their intrinsic unity, Professor often spoke of the Zarathustrian forces as Essene Angels. Actually, the Zarathustrian force of Food is included in Essene angelology as part of the Kingdom of the Earthly Mother.

a strong immediate faith through the evidence of intuition, by these hormonal enzymic factors. So he did teach them about air, water, food, everything they should know about how to change their eating and living habits, to result in permanent health. But he knew human nature and human weakness—he understood lack of understanding, lack of will power—so he said, now you are cured, but go, and sin no more. When sometimes a spiritual healing takes place, that is a wonderful thing—but if that person goes away the same ignorant person as before, and continues to sin, in the words of Jesus, he will be very soon sick again, and we didn't do him a very good service. Remember the old Chinese proverb: when you give fish to a hungry person, you helped him for a day. But if you teach him how to fish, you remedied his hunger for a lifetime. Now this is a very important thing. Somebody who lives and follows the Essene Way of Biogenic Living will not sin anymore—he will remain happy—he will get cured, naturally, by the law of cause and effect, because powerful biogenic forces will enter into action, all the catabolic waste products will be eliminated from his organism, he will breathe fresh air and eat good foods, and so on. And this is what I can tell you about healing in a nut shell. If you improve your eating and living and thinking habits, you will be permanently healthy, because you will sin no more. This is what Zarathustra and the *Essene Gospel of Peace* are teaching us.

This was the serious part I wanted to explain to you. Now comes the humorous part. In the twenties, spiritualism was a big vogue. There were people on street corners in Paris who had contact with Jesus, with John the Baptist, with Beethoven, with Genghis Khan, and I don't know who else. For ten or twenty francs they went into a trance and mumbled things and gave you some message. I remember once a good friend of mine told me, you know, you always quote Socrates—I said yes—he said, I know a medium who will put you in contact with Socrates—are you interested? Well, I have an open mind, and I am always glad to meet my old friend Socrates, so I said, let's go! So we went, and there was a rather sinister-looking person—very disturbed *tenue*— he insisted that first we pay twenty-five francs—and then he went into a trance and started to mumble and "Socrates" did speak, oh, for about twenty minutes. Then he came out of the trance and my friend asked me, well, what did you think about it?

I said, well, I read the dialogues of Socrates—tremendous, deep wisdom—but all these nonsenses I did hear from this person mean only two things: if he really did contact Socrates, then Socrates during the last twenty-five hundred years developed a serious case of senility, because what he mumbled about cannot be compared with what is in the dialogues of Socrates. And the other possibility is just that he is a crook, and wanted the twenty-five francs. So it was a fine circus. I had many friends who believed in mediums and they took me to Genghis Khan and John the Baptist and many others. I mean, I like circus, and in my free time it is wonderful entertainment. Now some people like to teach by praying together with you, or reading from the *Essene Gospel of Peace,* or beautiful spiritual literature like the *Zend Avesta* of Zarathustra, and you feel as if you are renewed. That is fine. But then there are those people who go around and say that for a modest fee they will put you in contact with and give you a message from some great saint or I don't know whom. Five per cent of these people have visions and hallucinations, which is evidently a pathological case needing medical treatment. Ten per cent of these people are big-sized crooks who are successfully hypnotizing a lot of people, collecting a lot of money with their mediumism. And about eighty per cent of them are small crooks who are doing it just for a little money because they are lazy to work or do some other honest thing. This is my experience with a great many cases in many countries.

Mediumism went out of mode in the twenties, but it still exists sporadically, here and there. I remember there was a great British scientist who believed in mediumism and finally decided to investigate it. Crookes was his name—he discovered thallium—a great physicist. And he dedicated the rest of his life to investigate mediumism and wrote a dramatic booklet describing how much he was disappointed—all the ignorance and deception he found in this field. Then, in the U.S. there was a very well-known person who investigated mediumism: Harry Houdini. Houdini was the greatest magician of all times—he made miracles, he materialized things before your eyes, and nobody could ever explain how he did these things. But he did them, simply as entertainment. When he decided that he would investigate mediumism, he also dedicated the rest of his life to unmask the crooks and deceivers who took away savings from old ladies and money from credulous people—because the majority of people are very gullible and believe these things, especially in an age of disorientation, like the present. He was very successful, too—he unmasked one famous medium after another. But that went out of fashion a long, long time ago, and it only seldom happens now, here and there.

So the conclusion is this: that spiritual healing is possible. But it is possible by your inner faith—a strong faith in the real forces that constantly surround us—this is a healing which is real and lasting, because it is based on reality.

*If a person were to do the Essene Communions very intensively for, say, three years—what would happen?*

Well, that's exactly what the Essenes did at the Dead Sea! And that's what the ancient followers of Zarathustra did—they lived with the Angels, in the language of the Essenes. When you do the communions, it changes you, it creates a new life. It gives us health, joy, peace—it makes a creative life for us. We achieve strength and longevity, and we are able to help our fellow man.

*What does it mean in evolutionary terms?*

In evolutionary terms it may mean that our future generations of descendants will inherit through phylogenetic heredity all these qualities, which is wonderful, and every generation will be better and better. Remember the expression, "the son of man." We find this very important expression in the Essene Gospel and other Essene writings. It means that when we are the embodiment of these sixteen natural and cosmic forces, visible and invisible, in that case we are no more that man of old that we were—we are renewed, and all our descendants will be renewed.

Norma: *Professor, what about the "antenna?" Isn't it our purpose also to grow an antenna?*

Well, now, that doesn't refer to a material antenna, or a radio antenna—but there is a kind of similarity. Like a radio antenna is receptive to certain currents, in the same way it is possible for us to become receptive—this is why in the individual inventory, the second question always deals with receptivity—do you *feel* these forces? That receptivity is extremely important—to transform yourself to develop a spiritual antenna. For instance, a great genius is an antenna—Beethoven had an antenna—he was able to draw from the music of the spheres, as mentioned by Pythagoras, the most beautiful things, in spite of the fact that he was deaf.

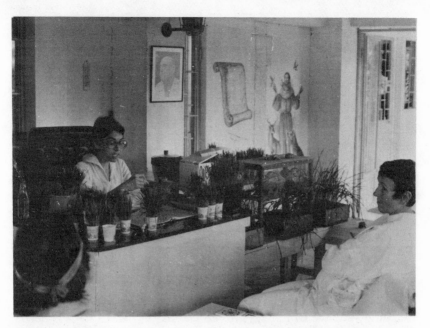

Socratic dialogue—questions and answers—was Professor's favorite way of imparting knowledge. Norma carries on the tradition at *(above)* the December 1979 Seminar and the July 1980 Seminar *(below,* with Sir George Trevelyan*)*.

According to Zarathustra and the Essenes, there is a Cosmic Ocean of Thought, and a Cosmic Ocean of Life—and a great poet or a great musician has a kind of antenna which can tune in to the most beautiful currents of universal thought and life. And there is a Cosmic Ocean of Love—this is why the Code of Ethic of Zarathustra was Good Thoughts, Good Words, Good Deeds. The totality of our cells, our vitality, is our Acting Body, according to the Essenes. The totality of our feelings is our Feeling Body. And the totality of our thoughts is our Thinking Body. Certain newly-developed methods of photography are able to show these to us. There is a solidarity of all thinking and feeling beings all over in the universe, and a genius can tune in to the Ocean of Love, or the Ocean of Life, or the Ocean of Thought—and then we are renewed. So it is worthwhile to work on our antenna—every day improving it and making it better and better. This is the main teaching of the Essenes and Zarathustra—we may not each of us be able to develop into a genius—absolute perfection may not be possible—but we can always go closer and closer to it. It is better to travel hopefully, than to arrive, said my old friend Lao Tzu.

*What is one sign during meditation that one is growing and developing?*

Now this is an interesting thing: an inner spiritual experience cannot be explained—those who have it, they experience it, but they don't talk about it. Those who talk about it, they don't know what it is. Buddha experienced Nirvana—but it is written in the Tripitaka, ". . . and the Illumined One did not explain it." Very simple. Somebody who is running around and telling all kinds of spiritual experiences he has, I assure you he doesn't know what he is talking about. Because those who have it do not talk—and when you will have it, you will experience it. It cannot be put into a definition or an explanation, because an inner spiritual experience is not in space—you cannot put your finger there and say it is at this point, here or there. It is not in time—sometimes a second feels like an eternity, and sometimes a long time feels like a second—it is not in time, it is not in space, it cannot be defined and it cannot be described for the same reason.

Norma: *I remember a beautiful poem of Kabir which says, "It is like a dumb person who has tasted a sweet thing; how can it be explained?*

Yes. Kabir was a poet-saint—the greatest poet of ancient India. His poems were translated by Rabindranath Tagore, the greatest poet of twentieth century India. It is a spiritual banquet to read the *Songs of Kabir.*

*Professor, I understand that grains were the last food to evolve on the planet, and some vegetarians feel it is the most perfect food for man. Grains have been traditionally the staple of many cultures and civilizations, and Zarathustra also praised grain— could you comment?*

Whole grains have a germ—like wheat germ, and others—which is rich in Vitamin E and Vitamin B complex; have polyunsaturated fats and excellent oil; have a lot of minerals—calcium, phosphorus, iron, copper, etc., and other nutrients of great value. And I always highly recommended their use in my books. But to limit ourselves to eat only grains, it will be one-sidedness—and as I always said, one-sidedness is the greatest mistake in life. I don't see any reason why you cannot eat sprouts, why you cannot eat biogenic greens, why you cannot eat fruits and vegetables and nuts—all these are wonderful things, wonderful biogenic and bioactive foods. Our nutrition should be balanced, and we should eat all these things and get the sources of a variety of vitamins, minerals, enzymes, hormones, and so on. There is no reason to limit ourselves to only one thing. But of course, a diet would be deficient without some whole grains.

*But isn't one of the problems of whole grains that they have to be cooked, or prepared in some manner?*

Not necessarily. For instance, sunflower seeds don't have to be cooked.

*But that is a seed.*

Well, grains are seeds! All grains and cereals belong to the seed family—they are seeds. Beside of wheat, perhaps the favorite plant of Zarathustra was the sunflower. It follows the sun. It is the only plant in the vegetable kingdom which has Vitamin D as well as some Vitamin E, excellent oil-soluble vitamins—it is tremendous. If you read *The Chemistry of Youth,* you will find a chart I give about the biochemical value of sunflower seeds, and you will recognize their value. Zarathustra never knew about the existence of this vitamin or that one—these are new discoveries—but through

201

observation, through experimentation, through intuition, through wisdom, he developed the sunflower seed, one of his greatest masterpieces. I consider that everyone should have a handful of sunflower seeds a day—it is an incredible storehouse of nutrients in the minimum of space.

*Professor, what about rice, millet, etc.?*

They are all fine whole grains. Remember, I recommended to eat 25% biogenic foods—we say, sprouts and very tender young greens—then 50% bioactive foods, which are fruits, vegetables, nuts, all these—and then you can afford 25% biostatic foods. Now if it is a question that we have a very serious disease and we need intensive therapeutic use, then we should use biogenic foods—it depends on the individual. For instance, you will not recommend to someone with a serious stomach ulcer to eat a big bowl of raw vegetable salad—he will be in the hospital the next day. Or if someone has a serious case of diabetes, you will not give him a bowl of dates and figs when the pancreas doesn't work well. *Est modus in rebus,* said my old Roman friends—there is a way in all things. Here is often one weak point of enthusiastic naturists and vegetarians—they make the mistake of not applying diet and nutrition to the individual case. No matter how excellent the diet, it must be applied to the individual case. No matter how fine is a food, it cannot be of benefit if it cannot be digested—and every individual is different. This is a very important thing: we must know physiology—we must know the human body, and it is worth a little study. Now you can advise people—there are certain universal validities—for instance, you can tell anyone with a clear conscience to stop eating white sugar, stop eating white flour, stop eating animal fat, stop drinking, stop smoking—now with that you have no risk—those recommendations will always help people. But if you give a food to somebody to eat which he cannot digest, it will not work. It will not work. So you have to be careful, you have to know the limits of where you can go. Then, when you enlarge your knowledge about the human body, through the study of physiology and biochemistry, then you can afford to make positive recommendations.

*Would you apply that formula of 25% biogenic, 50% bioactive, and 25% biostatic, to all environments?*

Yes, for general use—except if someone has some digestive

trouble, or some serious metabolic troubles that would prevent him from utilizing those 25% and 50%—that is a different story. But as a whole, generally, for daily living, it is an excellent system of nutrition, which I proved in practice with 123,000 patients I did treat during half a century. *Contra facta nihil valent argumenta*— against the facts, arguments have no value, said my old Roman friends. I speak about facts, empirical experience.

*Is it a good idea to clean the body out first, before starting the right diet?*

Yes, it is a good idea, but how to do it—that is sometimes a problem for people. You often see some enthusiastic naturists who have the idea, well, I will undergo a thorough cleansing. And they may end rather badly. Because I want you to realize that we are building millions of new cells every day. If you eat good, wholesome foods, you will build good, healthy cells. But we are also breaking down millions of old cells which are catabolic waste products being thrown continuously into the metabolism. So through exhalation of air, through perspiration, through solid bowel movements, through urination, etc., we are continually getting rid of catabolic waste products which are superfluous for the organism. You never will reach a point where you will say; I am 100% cleansed—I have no catabolic waste products. It just doesn't exist, because you are breaking down every day millions of dead cells and you will always have catabolic waste products. That is a physiological and biological law. Remember Socrates— we have to follow the path of reason and not imagine that now we go on and eat nothing but grapes for three weeks, or fast for twenty days, and we will be absolutely cleansed and cured! Not necessarily—because you are continuously breaking down old cells. And if you want to know, a person who fasts has more catabolic waste products in the system than a person who doesn't fast. And anyway, the idea is a fallacy that when you are fasting, you are not eating—because our most important food is air—you breathe twelve thousand quarts of air a day. Then another thing: when you fast, you are breaking down millions of fatty cells in the organism—you are on a pure fat diet when you are fasting. Oh, in certain cases it is useful—I wrote a book called *The Essene Science of Fasting*—where I advocated doing things correctly and in moderation—but I think we shall know human physiology—we must know the human body and what is going on in the organism.

*Professor, you have told us about all the great mystics and teachers who lived hundreds and thousands of years ago—but shouldn't we expect to progress and have new truths revealed to coming generations, or should we remain with the past eight thousand years of wisdom?*

Well, that is a very good question. The history of philosophy is nothing else but the history of different viewpoints in the face of the same reality. Now this is a very important sentence I wrote, oh, about fifty years ago. You see, the cosmos, the galaxy, our solar system, our planets, represent the same reality—things are changing continuously, but essentially it is the same. And this is interpreted by different philosophers through centuries and thousands of years. They have different viewpoints, which is fine, because it is nice to see different viewpoints—but the basic reality of the universe is the same. Matter in the cosmos is matter, energy is energy, life is life—and there are certain basic things, what we call eternal truth, which are eternal. This is why all the great teachings, separated by thousands of years, always teach essentially the same thing—because something is new does not mean it is necessarily better. I want you to realize that. Just because a book is published in the twentieth century does not make it better than a book of Plato, published twenty-five hundred years ago—the newness does not make it better. I always said, in talking about the "New Age," well, in this New Age we have thermonuclear missiles, we have wars and pollution and persecution—because our age is "new" doesn't make it better than previous ages. So when I speak about eight thousand years of wisdom—and I have a list in my book, *Books, Our Eternal Companions,* which gives the greatest works of the last eight thousand years—I not only include Zarathustra and Buddha and Jesus and the Essenes and the *Tao Te Ching* of Lao Tzu—I also include Kabir, I include Rabindranath Tagore, I include the great works of any age—it doesn't mean that we must limit ourselves to our age, because it is new. It will be an intellectual suicide to say that, oh no, is a new age, and whatever comes in this new age is the only exclusive truth, and all the great universal masterpieces are irrelevant because they were written long time ago and now we have something new. We have a tremendous number of new things which are very mediocre, very superfluous, and very harmful. So we have to use with caution the word "new." New doesn't mean necessarily better. Every year our technology, our industries come up with new models which you must buy because they are new, based on obsolescence. I want to tell you something—I had once a Cadillac built in 1928, and I still was using it thirty years later and it was better than all the new cars made at that time. It had practically no maintenance, no troubles, and was more efficient. That something new is better is an idea instilled into us so we will always buy new things—but we must not let ourselves be brainwashed.

*Professor, will there be a lot of earth changes taking place in the 1980's, as predicted?*

Well, I am always reminded of Nostradamus, who was considered one of the greatest predictors—I greatly appreciate one sentence in his predictions which reads: he who will not be in disease will be in good health. Prediction is worthwhile only if it is based on the law of cause and effect. If we predict something based on the observation of the cause, it necessarily shall bring that effect. So, for instance, earthquakes always did exist in history—there is no reason to think they will stop and not exist in the 1980's. It is just a simple observation of the law of cause and effect and Socratic reason.

*In a universe where all is the result of the motion of the interplay of two basic polarities, it seems paradoxical that one pole, light (good), be valued over another, darkness (evil). Are not both necessary to creation? What is the meaning and resolution of this apparent paradox?*

Very simple. You remember I explained to you the symbol of the Preserver—which later on after Zarathustra, some twenty-five hundred years later, became Yin and Yang of Chinese philosophy— it is the symbol that everything in the universe is correlative. There are always opposing things which complete each other, and one without the other will have no meaning. Everything is correlative—not just relative, as Einstein said, but correlative, according to Zarathustra. Sometimes one, sometimes the other is stronger. At daylight, we say the day is stronger than the night, because there is light and there is no darkness. Nevertheless, the darkness is coming at night—it is there. So each is an important ingredient of reality. Now for instance, you are in perfect health— it is evident that you are not sick. Nevertheless, one day you may become sick. And then you will be sick and not in good health.

The correlative, the opposite elements alternate with each other. When one is stronger, the other is weaker. When we strengthen our vitality and resistance, our enemy—disease—becomes weaker. When our enemy becomes stronger than our resistance and vitality, then we become sick—the enemy is stronger. So this is the law of correlativity. They both exist together, but at certain periods, certain cycles, one is stronger than the other. But the other always comes back.

*Professor, I have been told that the Quakers consider it a sin to have less than one year's supply of food stored at all times. Is this right, and how would you recommend storing wheat berries and other sproutables? Or how to preserve food?*

According to my view, the best storage is a producing vegetable garden and a producing orchard. They are continuously producing, you don't have to store anything. I don't see much virtue with storage, because with time, nutrients are getting lost. Of course, this is very easy in a climate like Costa Rica where you can have vegetables and fruits twelve months of the year. But it is not so easy in the north where you have five months of cold. Therefore, in the north, the solution is a greenhouse—a large greenhouse where you can maintain the right temperature and be able to produce what you need—here again the greatness in the smallness. In a greenhouse no larger than this terrace you can grow things continuously, as much as you need—all the tender greens and quick-growing vegetables.

*Have you revealed to us all of the esoteric knowledge known to the Essenes? If not, will you be doing so in future books?*

Well, I have an incorrigible disease: I always write new books. Nevertheless, my job is not to reveal to you everything, but to give you the method which you can use yourself to reach this knowledge by your individual efforts. Remember, Buddha didn't tell you what is reincarnation—he gave you a method of thinking—to harmonize in your mind the three universal laws, and so understand what is reincarnation. The same thing with Zarathustra—he gave you the method of individual inventory—how to get closer and closer to perfection. My job is to make you think. My most important work with the Seminar is to create intellectual fermentation—and make you think—in the spirit of Socrates, who said, I am just a midwife who is helping the truth to be born. Now

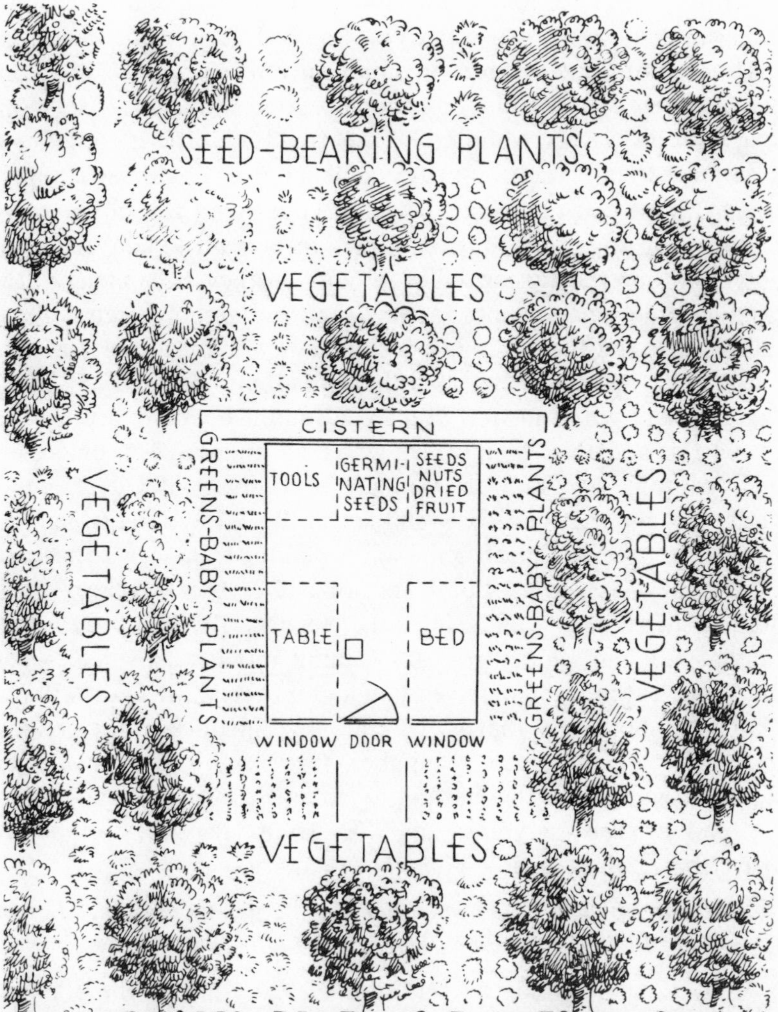
SEED-BEARING PLANTS
VEGETABLES
CISTERN
GREENS-BABY-PLANTS
TOOLS
GERMINATING SEEDS
SEEDS NUTS DRIED FRUIT
TABLE
BED
GREENS-BABY-PLANTS
VEGETABLES
VEGETABLES
WINDOW DOOR WINDOW
VEGETABLES
SEED-BEARING PLANTS
Essene orchard and garden according to the Plinius manuscripts in the scriptoria of the Benedictine monastery of Monte Cassino.

this is what I am trying to do—through Socratic dialogue to give you the method, the ancient and new traditions—everything I know to stimulate you to do your own thinking and reach your philosophy in your own way. You know, I never want to convert anybody from one religion or philosophy to another. If someone is a Christian, I want him to become a good Christian. If someone is a Buddhist, I want him to be a better Buddhist. If someone is a Zarathustrian, I want him to become a better Zarathustrian. And so on. I want you to become better in your own field, in your own philosophy, your own religion. But I never want to convert someone from one religion or philosophy to another. It happens all the time—people write to me: in view that you are an Essene, please tell me that and that—others write, in view you are a Zarathustrian, please tell me thus and so—others read my book, *The Living Buddha,* and write: in view you are a Buddhist, would you explain to me this and this—then others who read my books on precolumbian wisdom say: in view you are a follower of Quetzalcoatl, please tell me that and that—and so on. Well, the thing is, I am an Essene, I am a Zarathustrian, I am a Buddhist, I am an admirer of all the great teachings—but I am not one of them exclusively. It doesn't matter how wonderful is the teaching, if you limit yourself to that teaching alone, you did commit intellectual suicide—you did build a stone wall between you and universal life—you will not search anymore, and that is the end. One-sidedness is the greatest mistake in life. So please remember, this is my job—to make you think. I can give you the method to become a better whatever-you-are, and whatever is your "ism." Try to use the best methods to arrive there, wherever it is you want to arrive.

*Professor, please repeat the manner in which you feel Essene Living and Biogenic Eating can best be taught to others.*

Well, I may say in the tradition of Zarathustra, the simplest answer is the best: by example. *By example.* You see, if you live according to the Zarathustrian-Essene teachings, wherever you are, if you adopt Zarathustra's simplest ethical precept: *good thoughts, good words, good deeds*—this is the best example, in the field of ethics and in our contact with our fellow beings. This is such a simple code of ethics, even simpler than the Ten Commandments, simpler than the Roman Law, and simpler than the Code of Napoleon—simpler than anything. If we would practice this, we would never need even the United Nations. Zarathustra was a genius of simplicity, because this condenses everything: good thoughts, good words, good deeds.

Now you can also put into practice in your life the Cosmic Order. There is an ancient Spanish expression, from early medieval times: *donde está orden, hay Dios*—where there is order, there is God. Well, this is an intuitive reflection of Zarathustra's Cosmic Order. If you introduce into your daily life a certain order of priorities—because in life there are priorities—and if you don't go side-tracked with all kinds of ephemeral things, and don't become a dry leaf floating in the wind—but if you are purposeful—like the Avestic Creation—it was very purposeful, like everything Zarathustra taught—and if you concentrate on the essentials of life, for instance, trying to establish the maximum contact and absorbing the maximum of energy and harmony from those sixteen surrounding forces, doing your individual inventory and trying to put it into practice in your daily life—in that case it is impossible that your example will not have effect in your immediate environment. When, for instance, you have a small vegetable garden, you made a small orchard, you have a very small selected library, you did adopt the right method of thinking and recognize what is without value and ephemeral, giving preference and priority to the real values of life—when you eliminate from your life the wrong kind of pleasures, which are destroying your good health and peace of mind, and introduce instead the real pleasures of life, of nature, culture, and everything which is beautiful—in that case you will radiate in your environment all these wonderful forces, and in view of the general disorientation, people will invariably be curious about what you are doing and ask you questions—and there is no better, more fertile soil to transfer knowledge, than to answer the questions of curious people. They are thirsting for answers, for knowledge, because they don't have it. What they do have are substitutes—everything is substitute—we are living in an *ersatz* world. We substitute Bach, Mozart and Beethoven with all kinds of cacophonic music—electronic music which destroys your hearing. We are substituting the great masterpieces of all times with a tremendous number of newspapers and magazines and paperbacks and all kinds of worthless, ephemeral literature. Stop using substitutes, and go back to the original purity and simplicity of things. If you do that in your daily life, in that case remember what Romain Rolland said: the whole depends on the position of the atoms composing it. And then neighboring atoms will automatically borrow some of your methods and ways of living—and even if they adopt them only partially, it will make a change, and the change cannot start from governments above going downward, it has to start from the grass roots—(you see, we go back to the grass!)—and go upward. Because governments are not perfect—they are managed by human beings, and with governments there are always deficiencies, often corruption, and you cannot create a perfect society by an order from above. Unless it comes from the inside with every individual, and organically develops, nothing can be permanent. You may ask, well, how long may it take? Well, my answer is to be patient—give me a thousand years, give me ten thousand years—perfection is an ideal—we can only go closer and closer to it. If you try to utilize these sixteen forces of Zarathustra and the Essenes, then you will go closer and closer to it every day, and involuntarily you will lead closer and closer to it your fellow men around you. Because they will realize with time and experience and trial and error, that your ways are better and bring results—while the haphazard, erring round of different inferior influences does not lead them toward a more satisfactory life.

*Do you still feel that the time has not yet come for spiritual communities, and we should only attempt to have our own place in our own area with like-minded persons, rather than joining together?*

Well, this is an age when people are looking for all kinds of solutions—for the simple reason that they are not very happy with things as they are. Of course, by violating the laws of nature in physical health, as well as in peace of mind and in many other ways, naturally they are unhappy. They are looking for solutions, and it is human nature to look for short-cuts. Some people are looking for instant salvation, or instant cosmic consciousness, and in view of this, there are a great many efforts in the atmosphere in our age of changes. It is natural that when somebody becomes acquainted with some of the great teachings that they are yearning to live together with those people who similarly are trying to practice these teachings. So it happened, looking back on the last few decades, that communities are being formed. Now with these communities I did a lot of research, and I made a chart of some very interesting statistics, and only one out of twenty of these communities were successful. What does that mean? It does not mean that it is not possible—we shall not be discouraged—it means simply that there are certain preconditions—how a community based on natural and cosmic laws can survive and be successful: first, it depends on the degree of psychological maturity of the individuals composing it. Here comes again Romain Rolland: if the members of that community don't have that degree of emotional and psychological maturity, the community is doomed to failure because all kinds of emotional conflicts will develop, and with immaturity we cannot solve conflicts. When we face a problem, the best attitude is to use the symptoms of maturity—first, we shall use the principle of *prevention*—it is much easier to cure a disease by preventing it, than by trying to treat it afterward. The same way in a community—if members have the psychological and emotional maturity to foresee possible problems and troubles, they should use prevention—but only psychologically mature people can use prevention. Another important principle is *communication*. Mature people always are able to communicate their desires, their thoughts, their ideas. No community can exist in harmony unless its members practice communication. Then there is another principle, the principle of *compromise*. Nobody is in possession of the only and exclusive truth. When human beings live together—you may have your truth, and your friend may have his truth—and the real truth may be not yours, and not his, but somewhere halfway between the two. Therefore, unless we learned in life compromise, it is impossible for people to live together in a harmonious way.

Then there is another thing that usually happens. People are living in the city, working at a desk in an office somewhere, and they suddenly say, well, let's go somewhere and found a colony and start a new way of life! That's very beautiful—but if they don't have the knowledge and the skill, how to grow vegetables, how to plant and take care of trees, how to build a simple dwelling, how to use simple tools, and create the simple things in life that are necessary—well, it just will not work. I have told you about the summer seminars I gave in England in the thirties under the auspices of Sir Stafford Cripps—once he invited me to see him in his office, and as he was Chancellor of the Exchequer, always all kinds of people with fantastic ideas were coming to him, trying to persuade him to adopt this wild scheme or that utopian idea of how to manage the finances and economy of the British Empire. So he was sitting behind his desk, and there were a few chairs on the other side for people who had appointments to talk with him—and he had a little stick to the left of his desk, and behind him on the wall was a simple sentence in large letters: *In Life, Only Practical Things Work.* Then each time when somebody came with funny ideas which were not practical, he took that stick—he didn't even turn around—and just pointed to the wall behind him! And that was the end of the appointment. So I always think about him, that he was absolutely right, that in life only practical things work. They may be the most spiritual people who come together, but they will starve, they will freeze in winter, they will absolutely not be able to exist if they don't have the emotional, psychological and practical preconditions. During the last fifty years, I saw all kinds of colonies try to get started, and I remember many times they wrote to me for advice, and sometimes asking me to arbitrate in disputes—for example, there was a gentleman—I don't think any of you remember, it was a long time ago—Juan Wilkins, who founded a naturist colony on the Isle of Pine, south of Cuba—(at that time it was the old Cuba—Castro didn't exist yet)—the colony had about fifty people, and everyone got into everyone else's hair and they asked me to go and make peace—well, I kindly refused, because I had more important things to do. Oh yes, I was giving a seminar at Rio Corona in Mexico at that time, and didn't feel like traveling to Cuba.* Then there was another, Dr. Ziegmeister, who had a naturist colony in Chiriqui, in the Panama highlands—oh, it was a wonderful climate, fine preconditions—but there again was the same thing—a bunch of very spiritual and extremely impractical people—and everyone finally went back to the States and that was the end of it. And those were only two—I may say there were at least two hundred colonies and communities who wrote to me—once I made a statistic out of it and found that only one out of twenty was successful. Nevertheless, that one out of twenty was a success—why? One, because they believed in the same philosophy of living—extremely important—two, they were practical people—they knew gardening, they knew construction, they knew everything which was necessary to survive. Three, and most important, they had psychological

*Professor's adventure in the mahogany forest of Rio Corona is described in *Search for the Ageless, Volume One.*

maturity. This is why they were successful. But remember the statistic—one out of twenty.

Many years ago, some Quakers came to Costa Rica from the U.S. and settled near a rain forest at a high altitude—very wise choice of climate. They were somewhat fed up with the complicated technological way of living in the United States, fed up with the complex bureaucracy, the innumerable rules and regulations, the general pollution of the ecology, the diminishing of individual freedom, and so on. So they came to Costa Rica, a very nice group, and settled in that remote area called Monteverde. It was similar to the reasons why the Essenes chose the Dead Sea—they also wanted a good distance between them and the cities. Now the Quakers are wonderful people. Once I asked my friend Aldous Huxley what he thought about the Quakers, and he said, it is a tremendous relief to see people in permanent silence in a noisy world. For the Quakers pray in silence, nothing but silence. It is very interesting, that in the Cosmic Order of Zarathustra, in the *Zend Avesta,* the corresponding musical accord to the Creator—geometrically a point—is not a phrase of music, but silence. Now these Quakers are very hard-working, sober, practical people. They know how to build houses, they know how to create a unit of living wherever they go. And by introducing new and simpler, better methods in agriculture and ways of living, they exerted a tremendous influence on their neighbors. They taught the Costa Rican peasants how to do certain things in a much better and healthier way. As they settled that remote area, they created a wonderful atmosphere and a symbiosis and harmony with their environment. Yes, but they united there these three things, which are the basic essentials: belief in the same philosophy, practical ability, and psychological maturity. So they were successful. Now very often the I.B.S. receives letters—wonderful, idealistic young people who write, I would like to go to Costa Rica and live in the jungles off the fruits! Well, that's a wonderful idea—there are only a few problems involved with it: number one, I don't believe in tropical paradises—to do archeological field work there from time to time is one thing. But to live in the jungles is not a good idea. Low elevations, sea level in the tropics is a combination of heat and humidity which is very bad for your health. And also you have mosquitoes, you have snakes and all kinds of other harmful creatures, and it is definitely not an optimal area to live. So that is one thing. The other thing is, it doesn't matter what country you go to, you will find all over processed foods, chemicals, a certain amount of pollution, and other problems. Now the only way to avoid these things is to have an extremely small—maybe half an acre, or even a quarter of an acre—area where you can build yourself a simple dwelling, where you can plant a little vegetable garden and maybe a dozen trees, and if you are practical and know how to do things, and you understand the Cosmic Order and have the right way of thinking, and live according to the great teachings, you can have a wonderful life. But only you can do that You cannot afford others doing it for you, like in a community—it is only you who can do it. It is the individual who counts—it is the atoms which compose the whole, that count. Naturally, for this purpose you need a little capital to buy that half an acre, and to exist until your orchard and your vegetable garden will start to produce. And another thing—in every country there are some laws, immigration regulations, etc., and you have to satisfy the government that you will be independent. In every country there is a certain degree of unemployment, and they don't especially enjoy people coming there with nothing and becoming a charge on the government.

During the last decade, groups of hippies came to Costa Rica—they went straight to the beaches with knapsacks and sleeping bags and settled on those beaches and started to pollute them. They absolutely neglected hygiene and made a mess there. Well, the effect was that the Costa Rican government, which is very liberal and easy-going, had to use great persuasion to convince them to "please go somewhere else to do these things, because we appreciate our ecology and the beauty of our beaches." In fact, they created such an aversion that now when you go to a Costa Rican consul to get a visa to visit the country, if you go with long hair and a long beard, they will kindly suggest to you that you have a haircut and shave the beard—because they became allergic to those people who made a mess of their beaches, and they all had long hair and long beards.

Now that is just a humorous intermezzo—but I want you to realize that everything can be done if we use common sense, if we apply prevention, if we have the right method of thinking, and if we have the perseverance and the skill and the emotional maturity.

Of course, there is an ideal way to combine the philosophy of the community with the practicality of the individual homestead. We suppose there are a dozen people, or even a few dozen, who have similar ideas and ideals—fine. Each one can buy his own independent half an acre and they can be good neighbors and help each other. They can get together to study—have a communal library or something similar. This is what the ancient Essenes did— every member had a little dwelling which he owned, and a small orchard around it. If you read *Search for the Ageless, Volume II,* I think you will find there a description from the Plinius manuscript—even a graphic picture of a little Essene dwelling and the small area around it for intensive gardening, a few trees and plants and so on—all with the maximum simplicity. Then the Essenes got together every evening for study, for music, for spiritual experiences, for teaching, education—and that was a wonderful combination, because everyone was independent. You see, when you enter into a community, you are trying to escape from the complex laws and complications of a government, of a country. But what happens really is that you don't, because in addition you adopt new rules and regulations of that community. Wherever a group of people live together, there must be rules and regulations— otherwise there will be anarchy. Or something much worse can happen—instead of anarchy, you can have dictatorship—like, for instance, remember what was that place in Guyana—Jonestown— and several others, where people live under the dictatorship of a leader who is more immature than they are. Or you may have democracy, but with a lot of rules and regulations. Or people may give their money to a community, invest whatever they have, and then lose everything and have to leave because they cannot exist with all the rules and regulations.

Now instead of all that, in the spirit of the greatness in the smallness, if every person has a small place—we say, half or quarter of an acre, a little vegetable garden, a small orchard, a library, and builds a small dwelling—a simple, natural and spiritual life is possible, with absolute independence. And remember why you are independent—because you are the producer, and you are also the consumer.

In the United States at present, one person out of eight is working for the government—the federal government, the state government, or the municipal government—some kind of government. This means that every eighth person is pushing or filing papers in an office instead of going out and planting fruit trees or doing something productive. Well, that is big government—and what does big government want from the individual? Taxes. Higher and higher taxes. This is why there are revolts in many parts of the United States and in other countries against taxes. All right. Big governments want higher and higher taxes. Then big business— these international corporations—want bigger and bigger profits. Then big unions want higher and higher wages. And the individual is the slave of all the three. Everything you have to do, you have to pay income tax, sales tax, all kinds of taxes all the time to the government. Then everything you buy—all the superfluous things you are induced to buy through television, radio and press advertising, well, you have to pay a very high price—because that is the way the large corporations who manufacture these things make a lot of profit. And then, whenever your plumbing goes wrong and you ask your plumber to come, he will charge you fifteen dollars an hour to fix whatever it is—so you are exploited by the big unions, you are exploited by big business, you are exploited by big government.

Now if you have a subsistent creative health homestead, and you build a little dwelling for yourself, you grow your own fruits and vegetables, and you enjoy all the real pleasures of life—all that is beautiful in nature and culture—well, the government cannot ask very high taxes on a quarter or half an acre of land. It is no more a problem. Then, you will not buy a lot of superfluous things that you don't need and that are harmful for you—therefore, the large corporations cannot exploit you. And also, if you can do everything yourself that you have to do in such a small area, you don't have to ask from the unions to send you a plumber and pay an exorbitant price, because this is the *greatness in the smallness.* It is very easy to take care of a small place—very easy—you don't need more than two or three hours a day and you will have plenty of time to read, to enjoy good music and all the beautiful things— but if you have a large place, then you entered into the field of self-exploitation. You acquired a lot of things you don't need, and you have to pay with your free time, with your health, and with things which are really valuable for you. And you will get into a lot of troubles. Remember what Horatius said many thousands of years ago—my old friend Quintus Horatius Flaccus—the wind always breaks the tallest trees, but the modest little grass survives the tempest. And another old friend of mine, Diogenes, said a beautiful thing—that real freedom consists in the minimum of needs. Please meditate on that: *real freedom is in the minimum of needs.* Therefore, you don't have to look for your solution in complicated, large-size things like a community, with rules and regulations—where the community may become bankrupt and you can remain without anything. And also, you may lose your freedom—and freedom is the most beautiful thing and has the greatest value. And when you have a small, creative health homestead, I mean really small—half an acre maximum—in that case you can live a simple, natural, healthy life, and dedicate yourself to the real enjoyment of beautiful values of universal culture, and arts, and everything. Because please remember what I am telling you now: *capitalism and communism both have the same original sin: they both believe that the most important thing for the individual are material goods.* In the United States, the whole system is based on advertising—on radio, on television, in newspapers—to induce you to buy superfluous or harmful things. In Russia, they are not better—they are worshipping the tractor, and the machines—they are already developing pollution and the same troubles we have in western civilization—they are wasting their natural resources, very soon they will not have enough energy—they create large cities—what Thomas Jefferson called the "pestilential cities," where everything is polluted, and so on. I say it again: both capitalism and communism have the original sin: they say the most important thing for humans are material goods. Well, material things are becoming more and more limited on our planet because of the tremendous population growth—therefore, even with a perfect social system and distribution, there will be less and less material goods for each individual. But at the same time, there is an inexhaustible source of the real values and pleasures of life—good books, good music, spiritual experiences, peace of mind, good health—the enjoyment of everything which is beautiful in nature and culture—these are worth much more than material things we don't need—because there is very little that we really need. Even a small dwelling—in our guidebook, *The Essene Way— Biogenic Living*—we give an example of a biogenic dwelling which is extremely simple—anyone can build it. You only need a table, two chairs, a bed, a shelf for your books, and a good radio so you can get some fine music program—the real pleasures and values of life are simple—and that gives you free time and independence. Otherwise, you have a lot of troubles, you will have headaches, you will have nervous tension, you will have worries, you will not be in good health, and you will be part of a machine which is getting worn out. Because the economic systems of different countries are getting worn out. Every country wants to sell more to the other countries to have a good balance of payment. But it is evident—you don't have to be a genius economist to realize that every country cannot sell more to every country because the others have to buy it. So please remember, the only freedom consists in the minimum of needs.

But simplicity is not enough. If you consider only simplicity, you can build a shack, an ugly thing, and sit there and eat some canned foods—that is also simplicity, but it doesn't fill our need for beauty and culture. You have to understand the laws of nature, you have to understand the great teachings, you have to know how to live—but for this, you have to learn and learn and learn. Remember what Abe Lincoln said: I will go on and study, and my opportunity will come. Now people who first didn't acquire the right method of thinking, and the right practical knowledge and necessary skills, and are not mature psychologically—when they go out to some beautiful place to form a colony, they have a 95% chance to fail. So this is what I can tell you in a nutshell concerning these things.

*Professor, what have you discovered regarding the doctrine of reincarnation?*

Well, you know, when some complicated medical case comes to a general practitioner, then he usually refers you to a specialist. So, concerning your question, I have to do the same thing. I must transfer you to the greatest specialist in the field of reincarnation: Buddha—who was the greatest thinker of all times and who penetrated deeper into this matter than anyone else in universal history. Well, according to the ancient scripture of southern Buddhism, the Tripitaka, when a disciple asked Buddha what was Nirvana, he simply said, Nirvana is not existence; Nirvana is not non-existence; but if we say that Nirvana is not existence, nor non-existence, we still didn't define Nirvana—because it is something which we cannot define—it is an inner spiritual experience. And then the greatest Chinese philosopher, Lao Tzu, said, "He who knows, doesn't talk, and he who talks, doesn't know." Nevertheless,

Buddha, who was the greatest mystic of all times, was also the most scientific mystic of all times. Instead of giving to you a definition, he gave you a method—how you can arrive to the understanding of reincarnation yourself, by your own individual efforts. Well, Buddha said that there are three universal laws: the first is the law that *nothing is lost in the universe.* Now with this statement of Buddha, made in the sixth century B.C., we cannot dispute at the end of the twentieth century. Because we have exactly a scientific law which is proven and universally accepted, which is logical and evident: the scientific formulation is the law of preservation of matter and energy in the universe. It means that matter is continuously being transformed into energy, that energy is being continuously transformed into matter—but the sum total of energy and matter is the same in the universe. So contemporary science confirms the thesis of Buddha, which he formulated in a much simpler manner: nothing is lost in the universe.

The second universal law of Buddha is that *everything is continuously changing*—that there is not one solid point in the universe. Archimedes, the great Greek philosopher and geometrician, said: I wish I could find a solid point in the universe—from this solid point I could shake the whole cosmos. But there is no such thing. As Giordano Bruno said, every point in the universe is a central point, and everything is in continuous change and movement—in our body, our cells are continuously changing, our blood is continuously circulating, and on our planet there is the wind, there are the oceans, the rain, everything is in continuous change. In the cosmic space, the celestial bodies, the suns, the planets, the galaxies, are continuously moving and changing with tremendous speed. Everything is continuously changing. Long after Buddha, even the Greek philosopher Heraclitus said, *pantha rei kai horei*—which means in classic Greek, everything flows and counterflows. Therefore, we cannot dispute with this second universal law of Buddha, that everything is continuously changing.

And the third universal law of Buddha is that *all these changes are following the law of cause and effect.* Nothing is happening haphazardly. Everything has a cause and everything is creating an effect. This third law is the law of Karma. Therefore, if you can harmonize these three universal laws in your mind, said Buddha, then you will understand what is reincarnation. Now you realize that Buddha didn't give a definition of reincarnation; Buddha gave you a method how you can yourself reach to the understanding of reincarnation—and he gave you three tools, three universal laws. Well, I personally don't think I can improve on it, because nobody in universal history did analyze more profoundly the law of reincarnation, than Buddha.

*Can you speak about the significance of the following symbols: the dualism of the Preserver, the wheel of life of Eternal Life, the eight-pointed star of Work, the crescent moon of Peace, and the flower of Joy?*

Well, these five pictographs have their meanings—as do all the Zarathustrian symbols—I will explain them briefly. *(to Norma)* Now please put somewhere on the table so everyone can see, the symbol of the Preserver. This ancient Sumerian symbol was later on adopted in China, and in other cultures. It represents a very important principle—I will use a word which you will understand much better than if I use the ancient Sumerian expression—it means that the universe is correlative. Correlative means that, for instance, we have day and night—they complete each other and form one unit. We couldn't understand one without understanding the other. Like health and disease. Without one, the other would be unintelligible. So the ancient Sumerians conceived the whole universe as correlative—it consists of light and darkness, and so on. Everything is correlative, everything has its counterpart, without which the first would not be understandable, and they usually complete each other. So this is the symbol, and this cohesive effect, the affinity and the cooperation of these opposites which form a whole unit and complete each other—this it is which preserves the universe. This is the ancient symbol of the Preserver. Well, you can see it very well, it expresses the idea in its shape.

Now Eternal Life. Each of these symbols represents a step of the original Creation of the Zend Avesta which you saw outside on the platform. Well, this is a wheel with spokes in movement. You see, a wheel in continuous movement has no end—it is eternal and is in constant movement—life is movement. And this is the way the ancient Sumerians expressed the idea of Eternal Life. This, like all other ancient Zarathustrian and Sumerian symbols, did migrate through history in other civilizations, and in Tibet it became the Tibetan Wheel of Life. With Patanjali, the founder of the sixteen Yogas, it became the Wheel of Yogas, and so on. In fact, practically

220

each one of these most ancient Avestic symbols became, with a little change, a dominant symbol in other, later cultures.

Work. This is a very interesting cosmic concept and reveals the depth of wisdom in the *Zend Avesta* of Zarathustra. In the universe there is the cosmic space—or we may say the ether, or whatever word you want to use—and it is in movement, like everything else. And all the work of nature, of different living beings, etc., is going on in a celestial body, which is symbolized by a star. Our earth is one of them. Well, you can see that plants are growing, animals are moving and living their lives, rivers are flowing, oceans are moving, the wind, the rain, everything—all the work in the cosmos, in the universe, is being performed in celestial bodies, in the stars. And this is why work is a cosmic phenomenon, a cosmic privilege, and a cosmic right. This is why there is the beautiful statement: happy is the man who has found his task; he should not ask for any other blessing. That is a wonderful statement. And this is why they had for the symbol of Work, the star—because all the works of the infinite cosmos are going on in the celestial bodies.

Now the crescent moon of Peace. In the *Zend Avesta* of Zarathustra, the moon is the symbol of Peace. It is the symbol of Peace because the moon comes up at night when everything seems to sleep—the birds go to sleep, the metabolism of the trees and everything else slows down—and the rays of the moon are not active, not strong like the sun—but are subdued, and quiet. And if at night you look up at the moon, it in itself instills peace and inspires quietness. This is why the ancient Essenes greeted each other, peace be with you—and that comes from an Avestic source as well—the same greeting was used in ancient Sumeria. This is why the moon was used as a symbol of Peace.

The flower of Joy. Well, when you are in a flower garden—not in a painfully regimented flower garden—not in a Victorian flower garden—but a natural flower garden, where beautiful flowers grow spontaneously all over—you cannot help but feel joy. Here in Costa Rica, for instance, if you go out in nature, you will find hundreds of orchids, an orgy of orchids—and all kinds of flowers growing everywhere. There is nothing more joyous to look at than flowers—and to smell the fragrance of flowers. Flowers are an expression of beauty in nature. Remember that beautiful sentence in ancient Hindu and Persian literature: life is a sunbeam on a lotus leaf. The flower has always been associated with joy. And the

PRESERVER      ETERNAL LIFE

WORK

PEACE          JOY

fragrance of flowers has a tremendous effect on our olfactory sense—it has more effect on our state of mind than perhaps any other. Through universal history, the flower was always associated with an inner feeling of joy.

*Were the Essene gardeners following only the principles of Zarathustra, or had they received other teachings and influences, as well?*

Well, in the great majority the Essenes did follow the Zarathustrian principles of gardening, which is the most perfect system of gardening. Naturally, they had to adapt it to the harsh desert climate and to that excessive heat. They had to make some changes of their own which they found by the empirical method. For instance, they worked in two installments: they worked a little bit when the sun went down and the air became cooler, and they worked in the very early morning. Then during the middle of the day they rested and studied. So their way of living and their way of gardening had to change a little bit due to the different conditions and different climate and weather than Zarathustra had in ancient Persia. But the methods were the same—they adopted the same methods.

*Professor, you said that what we need in the twentieth century is Peace. The force that corresponds to Peace is Joy. How can we look at life so as to feel and express joy, when we constantly have above us dark clouds of destruction—chemicals, dirty air, inhumanity, cruelty, and human apathy? Advise us how to achieve joy in our lives, for if we can achieve more joy, then peace may come with it. . .*

There is one general condition which is a serious obstacle to joy, and that is a deep feeling of insecurity all over the world, a very deep feeling of insecurity. It doesn't matter who, poor or rich, or on what continent they are living—everything has become insecure toward the end of the twentieth century. Now there is one important thing I would like to tell you: at the end of the twentieth century, there is no such thing as absolute security—it doesn't exist. It doesn't exist, number one, because as I mentioned, it is the first time in history that mankind acquired the technological capacity to destroy this planet, and mankind along with it. Number two, the way as our technological civilization is going, we are exhausting our natural resources. We have now a kind of small hors-d'oeuvre to understand this fact: energy is becoming more and more scarce—with all the inevitable consequences. Then we are polluting our natural environment with unnatural speed, and also we have a very sick economic system in the world—tremendous inequality—eighty per cent of mankind live in sub-human dwellings, have no pure water, have deficiency diseases, lack of food, lack of hygiene, misery—in one word, we surely made a mess of our planet! And in view that immature human beings, who can fail and make errors of judgment, are handling all this vast arsenal of destructive weapons—(I glanced through the list of the arsenal of both Russia and the U.S. which figured in the SALT agreement—is eighty solid pages of enumeration of each category of weapons)—we are exposed to the possibility of a thermonuclear holocaust, we are exposed to revolutions, we are exposed to anarchy, we are exposed to dictatorships, we are exposed to general pollution, to exhaustion of natural resources—therefore, there is no such thing as absolute security at the end of the twentieth century. But please remember this: it is impossible to achieve absolute security, but we must learn the art of how to live happily with insecurity. Please remember that we can do a lot of things as individuals. You cannot change the world—it became too complex, too complicated for any individual to change the world instantly—that is pure utopia. A lot of people think that because they have the right philosophy, it is enough to tell people about it and then everyone will accept it and everything will be a paradise on earth—but that is pure utopia. And do you know what is the meaning of the word utopia? Nowhere. Well, the main thing is to learn the art of how to live happily with insecurity, and this is exactly what the science of Biogenics is teaching. For example, if you live in the city—you breathe twelve thousand quarts of air a day, and that air is absolutely polluted—so from your lungs your bloodstream carries these pollution by-products and distributes them to your cells. But if you follow the biogenic way of living, you can keep a few Biogenic Meadows in your room, and they will manufacture oxygen. You just go close to those biogenic leaves of grass and you will breathe the same wonderful air as you would in a forest or a meadow out in nature. In a little Biogenic Meadow there are hundreds of living things working to manufacture oxygen for you, and to create pure air. This is just a very small example—but this is why we have Biogenic Meditation, Biogenic Relaxation, Biogenic

Sleeping, Biogenic Breathing, Biogenic Dew Bath—all the many facets of Biogenic Living described in our guidebook, *The Essene Way—Biogenic Living*. How to achieve joy—it is the same basic idea—we may be able to do little about the general pollution in the city we live in, but we can do a lot to improve the climate of the room we live in, and sleep in, by having two Biogenic Meadows on both sides of our bed. We can change our micro-climate, that environment where we, as individuals, live and work. If you want to wait until the ecologist movement will achieve the cessation of pollution of air and water and soil and everything, you may have to wait a century. We don't have time to wait a century—we had better do something right now—and it shall be done individually—it cannot be done collectively. Instead of worrying about all the chemical additives in the foods in the markets, you can grow your own fruits and vegetables and avoid those chemical pollutants, and so on. There are a lot of things we can do, and that we shall do. It is no use to sit down and become gloomy and say, oh, the whole world is going to the dogs—that is a defeatist, negative attitude. What will happen to the world depends on us, on every individual—and when we do our duty to ourselves and to our fellow man, and to our environment—then we become active points in the universe. And then the whole, which depends on the atoms composing it, will be changed and improved. And concerning the possible worry about a thermonuclear holocaust, remember again my old friend Horatius, who said, *Et si fractus illabatur orbis, impavidum ferient ruinae*—which means in Latin that even if the whole world, everything will collapse, the ruins will find me— a fearless man—buried there. The most important thing, according to the stoic school of philosophy of ancient Greece, is *ataraxia*— the undisturbed peace of mind. The undisturbed peace of mind is faith in the cosmic order, and the feeling that we are individuals, doing our job to improve things, as much as it is possible. And nobody can do more. So remember my old friend, Epictetus, who said to have *ataraxia*—undisturbed peace of mind—and also my other old friend, Horatius. That is the way to achieve a meaningful, satisfactory way of life—by utilizing all these forces and radiating them in our immediate environment to help others around us, through example. The main thing to remember is that there is behind you eight thousand years of wisdom. Don't go side-tracked with some nonsenses and irrelevant things—use your priorities in life, and know what is essential and what is irrelevant.

*Professor, will there be an opportunity to meet with you personally if we want to become Teachers of Biogenic Living?*

Well, I usually from time immemorial always reserve the first day following the Seminar for individual appointments for those who want to talk with me personally. So on Wednesday—which is a very propitious day—according to the Greek-Roman mythology it is the day of Mercurius, who was the god of philosophers, vagabonds and travelers—on Wednesday, we say, from 10:30 on, I will be here and will be very glad to see whoever wants to talk with me personally. Give your names to Norma so she can make a chart, and we will have some semblance of cosmic order, and everything will be fine.

*And so ended the July 1979 International Seminar on the Essene Way and Biogenic Living. Professor had often said, with his usual sense of humor, that he "never intended to stay on this planet forever." Two weeks after this last session, he died. His last words were exactly the ones he used to close this Seminar: "Everything will be fine."*

*And so it will be. Edmond Bordeaux Szekely did not believe in death—not for the individual, and not for the planet. No matter how dismal the outlook seemed for this world we live in at the end of the twentieth century, he always kept intact his "incorrigible optimism." He had absolute faith in the shining determination and undaunted perseverance of the individual—strengthened by the reality of the ageless teachings of eight thousand years of wisdom.*

*We must remember most of all that faith of his—and make him proud of us with our joyous and never-ending efforts. And may we also never lose the memory of that unforgettable twinkle in his eye—the joy of living that overflowed his being even during the most solemn discourse.*

*Often, when asked to advise the individual how he could best serve humanity during these troubled times, Professor would quote the last words of Buddha: "Strive incessantly." Let us make Professor's last words come true by joyously and actively following the words of Buddha. Nothing would make him happier.*

# APPENDIX

## QUESTIONNAIRE FOR INDIVIDUAL INVENTORY

(from *The Essene Book of Asha,* by Edmond Bordeaux Szekely)

Each of the sixteen items in the questionnaire is to be considered from three aspects. First: Is the power or force thoroughly *understood?* Second: Does the individual *feel* the importance and significance of it deeply and sincerely? Third: Is the power or force *used* continually?

*I.* **SUN**   The sun is a very important source of energy which man should utilize to the utmost.

   a) Do you understand completely the function of solar energy in your organism?

   b) Do you know in what way you can contact and utilize this energy?

   c) Do you utilize it in the form that is best for your health and well-being?

*II.* **WATER**   Everyone should have a bath each morning and use water in the optimal way in diet.

   a) Do you understand the influence and effect of water upon your health?

   b) Do you feel deeply the role of water in your life?

   c) Do you utilize water in the most efficient ways and do you take a bath every day of the year?

*III.* **AIR**   The importance of spending as much time as possible outdoors and of breathing fresh air.

   a) Do you know the role of air in life and its effects on the body?

   b) Do you feel deeply the need of pure air and right breathing?

   c) What practical measures do you take to utilize the energy of the atmosphere for your health?

*IV.* **FOOD**   Food of the right kind and in the right amount supplies another vital force of the human organism.

   a) Do you know the influence and effect of food on health? And do you know what are the best foods for the human organism?

   b) Do you feel deeply the importance of the right food for your own well-being?

   c) Do you utilize this knowledge and feeling in the best way, and do you practice what you know continually?

*V.* **MAN**   This represents each person's duty toward his own individual evolution. Man has to try to utilize every moment to further his progress in life, and it is a job which no one can do for him. Nobody and nothing can relieve a man of this right and responsibility.

a) Do you know and understand all your potentialities and do you know the most practical way of transforming them into reality?

b) Do you feel deeply the importance and necessity of developing your latent capacities?

c) What are you actually doing to bring out the inner aptitudes you have and to develop them to the highest extent so that you may progress little by little in your individual evolution?

*VI.* **EARTH**  Earth represents two aspects of generative force: that which creates life from the soil in the shape of plants, trees, grass and flowers; and that in the human organism which manifests itself in sexual life. Both aspects of generative power create more abundant life on this planet.

1.  a) Do you know the best and most practical way of growing plants and food, and the importance of so doing for your own health and the well-being of mankind?

b) Do you feel deeply the urge to grow things and the need for so doing?

c) Do you in fact grow things and make every effort to get the opportunity to do so?

2.  a) Do you understand what are the optimal ways of harmonious sexual life and do you realize its importance for your physical and mental health?

b) Do you feel this deeply within you?

c) Do you practice complete harmony with and obedience to the Law in this respect?

*VII.* **HEALTH**  This signifies man's relationship to all preceding forces: Sun, Water, Air, Food, Man and Earth; as well as the following one, Joy, for harmony with all of these is necessary for optimal health.

a) Do you realize and understand the importance of good health and of thinking, feeling and acting in the ways most conducive to health?

b) Do you feel deeply the need for health for your own sake and for others?

c) What are you actually doing to improve your health in respect to all factors which influence it?

*VIII.* **JOY**  It is man's essential duty and right to be joyous at all times, and when he is in the service of the Law he is always joyous and happy. Sadness and joy have a deep influence on those around us; both are contagious emotions. It is therefore of utmost importance that man should radiate joy at all times.

a) Do you understand the importance of joy in life, and on the health and happiness of those around you?

b) Do you feel the joy of living surging within you continually and do you feel its creative force radiating around you?

c) Do you perform all your daily activities with this deep feeling of joy and do you try to spread this joy around you?

These eight groups of questions represent the visible forces of Nature. The following eight questions concern the invisible powers of the Cosmos, which are even more important. Man lives in the midst of these cosmic powers and natural forces and he cannot separate himself from them even for an instant. That is why it is so essential that he shall strive continually to create a positive attitude and live in harmony with all of them.

*IX.* **POWER**    Power is manifested continually through man's actions and deeds. These are the results of his cooperation with all the other powers and forces in accordance with the iron law of cause and effect. He can only perform good deeds when he is in harmony with all the other laws.

a) Do you understand the importance of good deeds and that your personality, position and environment in life is the result of your past deeds, and that your future will be exactly what your present deeds make it?

b) Do you feel deeply the necessity of performing good deeds at all times and do you consider it a right as well as a duty?

c) Do you actually perform good deeds, and do they express harmony with all the laws of nature and the Cosmos?

*X.* **LOVE**    Love is manifested in the form of good words, for words reveal our attitude to other people. If we feel love toward others, we will speak only gentle and kind words.

a) Do you understand the importance of good words in your own health and emotional life and for the health and well-being of your fellow man?

b) Do you feel a sincere love for all beings around you?

c) Do you practice this diet of harmonious feelings and words towards all beings at all times?

*XI.* **WISDOM**    This represents man's duty to increase his knowledge and understanding in every possible way and to utilize every source of knowledge so that he may improve not only its quantity but its quality. Wisdom is manifested in the form of good thoughts. An individual may be clever and have great store of knowledge, but without good thoughts he has no wisdom.

a) Do you understand the extreme importance of a good diet in thoughts for your own health? And do you understand the value of increasing your wisdom so that you may always have good thoughts?

b) Do you long for true wisdom with a deep inner urge?

c) Do you grasp every opportunity to advance and grow in wisdom so as to understand more and more the cosmic order and your role in it? And do you hold only good thoughts in your consciousness and refuse even to entertain negative, destructive, thoughts about any person, place, condition, or thing?

*XII.* **PRESERVATION OF VALUES**  This power concerns the preservation of all useful things and good values. When anyone destroys a good thing or lets it be spoiled or damaged he is cooperating with the negative, destructive forces of the world. He must use every opportunity to prevent damage to whatever has value, whether a tree, plant, house, relationship between people, or harmony in any form.

    a) Do you realize the importance of preserving all good things, both material and immaterial?

    b) Do you feel deeply the need to conserve everything possible and to let nothing deteriorate or go to waste?

    c) Do you practice conservation in every way possible and at all times?

*XIII.* **CREATION**  This signifies the necessity for man to utilize his creative powers; his role on this planet is to continue the work of the Creator. He must therefore try to do something original and creative, something new and different, as often as he can. Every invitation to the Creator to work with and within him, strengthens man's creative power.

    a) Do you understand the importance of doing or making something creative or original so that you may truly cooperate with the Creator?

    b) Do you feel deeply the need for doing something creative and realize the inner satisfaction which results from it?

    c) What do you actually do that is original and creative?

*XIV.* **ETERNAL LIFE**  This concerns man's sincerity in all he does and with all those he meets. It concerns also his sincerity with himself and in answering all the questions of this questionnaire.

    a) Do you understand the necessity of being sincere with all people, including yourself?

    b) Do you feel a deep sincerity in analyzing your relationship with all sixteen forces and powers of nature and the Cosmos?

    c) Can you accept yourself as you are, without rationalizing to justify things you are doing, saying, or thinking?

*XV.* **WORK**  Work is the precondition of many other values. Work means, among other things, the performance of our daily tasks, whatever they may be, with honesty and efficiency. Work is man's contribution to society and is a precondition of happiness for all concerned. For when one person does not perform his work properly, others have to do it.

    a) Do you realize the importance of work and the necessity for doing your particular work with care and sincerity, both for your own sake and for that of your social environment?

    b) Do you have a deep feeling of satisfaction in your work?

    c) Do you do all your daily tasks efficiently and conscientiously and so return to society all you receive from it?

*XVI.* **PEACE**  It is man's function and duty first of all to maintain inner peace and create it within himself, and then to create and maintain it around him. He should never lose an opportunity to establish peace wherever he finds it lacking. If he will do this in his environment, and within himself, he will be helping to prevent inharmony, enmity and wars, since the condition of the whole of humanity depends upon the condition of its atoms—individual men.

> a) Do you realize the importance of peace, both within and around you, for the maintenance of your own health and the happiness of others?
>
> b) Do you feel deeply the need for this inner peace?
>
> c) Do you possess this inner peace and are you trying to do all you can to maintain it within yourself and to establish and maintain it wherever you are?

*St. Francis and the Animals*—life-size painting on a wall of the semi-outdoor lecture hall of the I.B.S. International Correspondence Center in Orosi, Costa Rica, by Antonielena Martínez, the talented Mexican artist responsible for many of the illustrations in Dr. Szekely's books, including the reproductions in this volume from *El Juego de los Dioses.*

# CREDO

*of the International Biogenic Society*

We believe that our most precious possession is Life.

We believe we shall mobilize all the forces of Life against the forces of death.

We believe that mutual understanding leads toward mutual cooperation; that mutual cooperation leads toward Peace; and that Peace is the only way of survival for mankind.

We believe that we shall preserve instead of waste our natural resources, which are the heritage of our children.

We believe that we shall avoid the pollution of our air, water, and soil, the basic preconditions of Life.

We believe we shall preserve the vegetation of our planet: the humble grass which came fifty million years ago, and the majestic trees which came twenty million years ago, to prepare our planet for mankind.

We believe we shall eat only fresh, pure, natural, whole foods, without chemicals and artificial processing.

We believe we shall live a simple, natural, creative life, absorbing all the sources of energy, harmony and knowledge, in and around us.

We believe that the improvement of life and mankind on our planet must start with individual efforts, as the whole depends on the atoms composing it.

We believe in the Fatherhood of God, the Motherhood of Nature, and the Brotherhood of Man.

---

☐ I feel an affinity with your Credo. Please mail to me your introductory booklet, with a complete list of I.B.S. publications.

☐ I want to study the complete, all-comprehensive guidebook and encyclopedia, *The Essene Way—Biogenic Living.* I enclose my check in U.S. currency of $9.68 ($8.80, plus 10% postage and handling), made out to **I.B.S. Internacional.**

☐ I want to become an Associate Member of the *International Biogenic Society.* My membership entitles me to receive free: (1) the comprehensive encyclopedia of ancient wisdom and modern practice, *The Essene Way—Biogenic Living,* (2) The Periodical Review of Biogenic Living, *The Essene Way,* (3) 20% discount on all publications of the I.B.S., (4) my membership card with all privileges, and (5) periodical publications. I enclose my membership fee of $20 (check in U.S. currency made out to **I.B.S. Internacional**).

☐ I want to receive the announcement and program of your *International Summer and Winter Seminars* on *The Essene Way* and *Biogenic Living,* held at the beautiful International Correspondence Center of the I.B.S. in Orosi, Costa Rica.

Please check the appropriate boxes and return this page to the International Biogenic Society, mailing address: **I.B.S. INTERNACIONAL, APARTADO 372, CARTAGO, COSTA RICA, CENTRAL AMERICA.** *(Air Mail Only!)*

*Name*_____

*Address*_____

*City*_____*State*_____*Zip*_____

234

# AN INVITATION

## to the I.B.S. International Center in Costa Rica

Over the years, thousands upon thousands of truth-seekers from all over the world have written to us, asking for the opportunity to study and put into practice the Essene Biogenic teachings in a harmonious framework, in the company of those motivated by similar ideas and ideals. In response to this need, the International Biogenic Society, in cooperation with Norma Nilsson Bordeaux, successor of Edmond Bordeaux Szekely, conducts every December an *International Seminar on The Essene Way and Biogenic Living,* attended by members, students and teachers from all over the world.

The Seminars take place at the beautiful International Center of the I.B.S. in Costa Rica, where participants study the ageless truths of the Essene Way of Biogenic Living from a semi-outdoor lecture hall overlooking seven mountains and four rivers, with the legendary volcanic peaks of Irazu and Turrialba towering in the distance, surrounded by lush green vegetation and the scent of tropical flowers. But physical beauty is not the only reason for the choice of Costa Rica as the site of the International Center of the I.B.S. As those who have read Edmond Bordeaux Szekely's *The Greatness in the Smallness*\* know, Costa Rica is the home of the World University of Peace—where pacifism is a way of life, where there is no army or manufacture of arms, an oasis of democracy and enlightened freedom in an increasingly troubled world.

If you would like to receive the announcement and program of the *International Seminars on The Essene Way and Biogenic Living,* please write to **I.B.S. Internacional, Apartado 372, Cartago, Costa Rica, Central America,** by *Air Mail* only.

\*available from the above address for $7.50, plus 10% postage and handling.

The volcano of Irazu slumbers at a comfortable distance as Norma describes the cosmology of Zarathustra at the December 1979 Seminar in Costa Rica.

*Reservation Form*

*Please return this form to* **I.B.S. INTERNACIONAL, APARTADO 372, CARTAGO, COSTA RICA, CENTRAL AMERICA.** *Use Air Mail Only!*

*Please enroll me in the:*

_____Winter Seminar on *The Essene Way* (December 26, 27, 28)

_____Winter Seminar on *Biogenic Living* (December 29, 30, 31)

I enclose $_____ (Tuition fee is $50 per person for each three-day Seminar)

*Name*_____

*Address*_____

*City*_____*State*_____*Zip*_____

Please send tuition fee to the above address in the form of a Cashier's Check in U.S. currency made out to **I.B.S. Internacional,** always using *Air Mail* (for safety, we suggest using *Registered Air Mail*). Please write clearly the full names and addresses of those coming with you, and remember to include tuition fee for each person. We look forward to welcoming you!

# RECOMMENDED BOOKS FOR STUDY

Many members who cannot attend the International Seminars because of distance or limited means are interested in a systematic program of home study. The following books are recommended for such a program, and provide an excellent foundation for study of *The Essene Way and Biogenic Living*, especially when coordinated with the methods outlined in *The Art of Study: the Sorbonne Method*. Also, it is recommended that these books be read and studied before attending the *International Seminars on The Essene Way and Biogenic Living*, held every winter at the I.B.S. International Correspondence Center in Orosi, Costa Rica.

Please send me the following books:

_____*The Essene Gospel of Peace, Book One* $1.00

_____*The Essene Gospel of Peace, Book Two* $5.80

_____*The Essene Gospel of Peace, Book Three* $5.60

_____*The Essene Way—Biogenic Living* $8.80

_____*The Chemistry of Youth* $7.50

_____*From Enoch to the Dead Sea Scrolls* $4.80

_____*The Essene Code of Life* $3.50

_____*The Essene Science of Life* $3.50

_____*Archeosophy, a New Science* $4.80

_____*Books, Our Eternal Companions* $3.50

_____*Cosmos, Man and Society* $5.80

_____*Search for the Ageless, Volume I* $7.80

_____*The Ecological Health Garden, the Book of Survival* $4.50

_____*The Book of Living Foods* $3.50

_____*The Greatness in the Smallness* $7.50

_____*The Art of Study: the Sorbonne Method* $3.50

_____Total for books

_____Less applicable discount*

_____Plus 10% for postage & handling

_____Total amount enclosed

*Please address all orders (by Air Mail) to:*
### I.B.S. INTERNACIONAL
**Apartado 372, Cartago, Costa Rica, Central America**

Name_____

Address_____

City_____State_____Zip_____

## *Important Ordering Information

All orders must be prepaid. Be sure to use Air Mail only. Minimum order: $5.00, minimum postage: 75¢. Please make Cashier's Check or Money Order in U.S. currency out to *I.B.S. Internacional*. Allow four weeks for processing. Members of the International Biogenic Society may deduct 20% and *active* Teachers of Biogenic Living may deduct 40% (both must have valid membership cards for the current year). Dealers and distributors, please write to above address for discount information. All sales are considered final.

## APPLICATION FOR ASSOCIATE MEMBERSHIP
## INTERNATIONAL BIOGENIC SOCIETY

**Please return to:**   *I.B.S. INTERNACIONAL*
*Apartado 372*
*Cartago, Costa Rica*
*Central America*

*Date_____*

*Name_____*

*Address_____*

*City_____State_____Zip_____*

*Age_____Profession_____*

*Previous Experience_____*

*I am interested in____becoming an Associate Member of the I.B.S.*

*____becoming a Teacher of Biogenic Living.*

Enclosed is my annual Associate Membership fee of $20.00. Please mail my membership card, your next issue of the Periodical Review, and my copy of *The Essene Way—Biogenic Living,* my most important textbook and encyclopedia of ancient wisdom and modern practice. I understand I will receive a 20% discount on all publications as an Associate Member, but only if I order *directly* from I.B.S. Internacional.

*Please make your check in U.S. currency out to*
*I.B.S. INTERNACIONAL*

Please use *Air Mail* only.

"Let us leave on one side all that divides us, all our shades of thought—political, social, religious, philosophical! We are not concerned to work out a single doctrine to be laid before the world for approval. Every doctrine—be it scientific or religious—is subject to discussion. In its anxiety to bring about a unity of minds, doctrine destroys what it seeks to establish.

"We are concerned, at this present hour, to build, over the whole world, a single front against war, violence, egotism, ignorance. Let us ordain resistance, opposition, refusal—an unqualified 'No!'—to war.

"And, if we need a central principle on which to rest our action, let that suffice which consists in solidarity and mutual aid, shall I say, in *communion between all living beings*. Let us unite all the spiritual forces of Life against the forces of death."

—Romain Rolland: Preface to *La Vie Biogenique,* 1928.

O that my words
Were graven
With an iron pen
In the rock
Forever!
For I know
That my Creator
Liveth:
And he shall stand
At the end of time
Upon the earth
And the stars.
And though worms
Destroy this body
Yet shall I see God.

—*Essene Gospel
of Peace, Book III*